MATHEMATICAL DISCOVERY

GEORGE POLYA
PROFESSOR EMERITUS OF MATHEMATICS, STANFORD UNIVERSITY

JOHN WILEY & SONS, INC.

MATHEMATICAL

On understanding, learning, and teaching problem solving

DISCOVERY VOLUME I

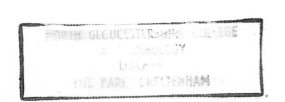

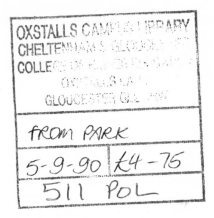
10 9 8 7 6

LIBRARY OF CONGRESS CATALOG CARD NUMBER: 62-8784
PRINTED IN THE UNITED STATES OF AMERICA
ISBN 0 471 69333 2

PREFACE

1. Solving a problem means finding a way out of a difficulty, a way around an obstacle, attaining an aim which was not immediately attainable. Solving problems is the specific achievement of intelligence, and intelligence is the specific gift of mankind: solving problems can be regarded as the most characteristically human activity. The aim of this work is to understand this activity, to propose means to teach it, and, eventually, to improve the problem-solving ability of the reader.

2. This work consists of two parts; let me characterize briefly the role of these two parts.

Solving problems is a practical art, like swimming, or skiing, or playing the piano: you can learn it only by imitation and practice. This book cannot offer you a magic key that opens all the doors and solves all the problems, but it offers you good examples for imitation and many opportunities for practice: if you wish to learn swimming you have to go into the water, and if you wish to become a problem solver you have to solve problems.

If you wish to derive the most profit from your effort, look out for such features of the problem at hand as may be useful in handling the problems to come. A solution that you have obtained by your own effort or one that you have read or heard, but have followed with real interest and insight, may become a *pattern* for you, a model that you can imitate with advantage in solving similar problems. The aim of Part One is to familiarize you with a few useful patterns.

It may be easy to imitate the solution of a problem when solving a closely similar problem; such imitation may be more difficult or scarcely possible if the similarity is not so close. Yet there is a deep-seated human desire for more: for some device, free of limitations, that could solve all

v

problems. This desire may remain obscure in many of us, but it becomes manifest in a few fairy tales and in the writings of a few philosophers. You may remember the tale about the magic word that opens all the doors. Descartes meditated upon a universal method for solving all problems, and Leibnitz very clearly formulated the idea of a perfect method. Yet the quest for a universal perfect method has no more succeeded than did the quest for the philosopher's stone which was supposed to change base metals into gold; there are great dreams that must remain dreams. Nevertheless, such unattainable ideals may influence people: nobody has attained the North Star, but many have found the right way by looking at it. This book cannot offer you (and no book will ever be able to offer you) a universal perfect method for solving problems, but even a few small steps toward that unattainable ideal may clarify your mind and improve your problem-solving ability. Part Two outlines some such steps.

3. I wish to call *heuristic* the study that the present work attempts, the study of means and methods of problem solving. The term heuristic, which was used by some philosophers in the past, is half-forgotten and half-discredited nowadays, but I am not afraid to use it.

In fact, most of the time the present work offers a down-to-earth practical aspect of heuristic: I am trying, by all the means at my disposal, to entice the reader to do problems and to think about the means and methods he uses in doing them.

In most of the following chapters, the greater part of the text is devoted to the broad presentation of the solution of a few problems. The presentation may appear too broad to a mathematician who is not interested in methodical points. In fact, what is presented here are not merely solutions but *case histories* of solutions. Such a case history describes the sequence of essential steps by which the solution has been eventually discovered, and tries to disclose the motives and attitudes prompting these steps. The aim of such a careful description of a particular case is to suggest some general advice, or pattern, which may guide the reader in similar situations. The explicit formulation of such advice or such a pattern is usually reserved for a separate section, although tentative first formulations may be interspersed between the incidents of the case history.

Each chapter is followed by examples and comments. The reader who does the examples has an opportunity to apply, clarify, and amplify the methodical remarks offered in the text of the chapter. The comments interspersed between the examples give extensions, more technical or more subtle points, or incidental remarks.

How far I have succeeded I cannot know, but I have certainly tried hard to enlist the reader's participation. I have tried to fix on the printed page whatever modes of oral presentation I found most effective in my classes.

By the case histories, I have tried to familiarize the reader with the atmosphere of research.   By the choice, formulation, and disposition of the proposed problems (formulation and disposition are much more important and cost me much more labor than the uninitiated could imagine) I have tried to challenge the reader, awake his curiosity and initiative, and give him ample opportunity to face a variety of research situations.

4. This book deals most of the time with mathematical problems. Nonmathematical problems are rarely mentioned, but they are always present in the background.   In fact, I have carefully taken them into consideration and have tried to treat mathematical problems in a way that sheds light on the treatment of nonmathematical problems whenever possible.

This book deals most of the time with elementary mathematical problems.   More advanced mathematical problems, however, although seldom referred to, led me to the conception of the material included.   In fact, my main source was my own research, and my treatment of many an elementary problem mirrors my experience with advanced problems which could not be included in this book.

5. This book combines its theoretical aim, the study of heuristics, with a concrete, urgent, practical aim: to improve the preparation of high school mathematics teachers.

I have had excellent opportunity to make observations and form opinions on the preparation of high school mathematics teachers, for all my classes have been devoted to such teachers in the last few years.   I hope to be a comparatively unprejudiced observer, and as such I can have but one opinion: *the preparation of high school mathematics teachers is insufficient.* Furthermore, I think that all responsible organizations must share the blame, and that especially both the schools of education and the departments of mathematics in the colleges should very carefully revise their offerings to teachers if they wish to improve the present situation.

What courses should the colleges offer to prospective high school teachers? We cannot reasonably answer this question, unless we first answer the related question: *What should the high schools offer to their students?*

Yet this question is of little help, you may think, because it is too controversial; it seems impossible to give an answer that would command sufficient consensus.   This is unfortunately so; but there is an aspect of this question about which at least the experts may agree.

Our knowledge about any subject consists of *information* and of *knowhow.*   If you have genuine *bona fide* experience of mathematical work on any level, elementary or advanced, there will be no doubt in your mind that, in mathematics, know-how is much more important than mere possession of information.   Therefore, in the high school, as on any other level, we should impart, along with a certain amount of information, a certain degree of *know-how* to the student.

What is know-how in mathematics? The ability to solve problems— not merely routine problems but problems requiring some degree of independence, judgment, originality, creativity. Therefore, the first and foremost duty of the high school in teaching mathematics is to emphasize *methodical work in problem solving.* This is my conviction; you may not go along with it all the way, but I assume that you agree that problem solving deserves some emphasis—and this will do for the present.

The teacher should know what he is supposed to teach. He should show his students how to solve problems—but if he does not know, how can he show them? The teacher should develop his students' know-how, their ability to reason; he should recognize and encourage creative thinking—but the curriculum he went through paid insufficient attention to his mastery of the subject matter and no attention at all to his know-how, to his ability to reason, to his ability to solve problems, to his creative thinking. Here is, in my opinion, the worst gap in the present preparation of high school mathematics teachers.

To fill this gap, the teachers' curriculum should make room for *creative work on an appropriate level.* I attempted to give opportunity for such work by conducting seminars in problem solving. The present work contains the material I collected for my seminars and directions to use it; see the " Hints to Teachers, and to Teachers of Teachers " at the end of this volume, pp. 209–212. This will, I hope, help to improve the mathematics teacher's preparation; at any rate, this is the practical aim of the present work.

I believe that constant attention to both aims mentioned, the theoretical and the practical, made me write a better book. I believe too that there is no conflict between the interests of the various prospective readers (some concerned with problem solving in general, others with improving their own ability, and still others with improving the ability of their students). What matters to one type of reader has a good chance to be of consequence to the others.

6. The present work is the continuation of two earlier ones, *How to Solve It* and *Mathematics and Plausible Reasoning*; the two volumes of the latter have separate titles: *Induction and Analogy in Mathematics* (vol. 1) and *Patterns of Plausible Inference* (vol. 2). These books complete each other without essential overlapping. A topic considered in one may be reconsidered in another, but then the treatment is different: other examples, other details, or other aspects are offered. And so it does not matter much which one is read first and which one is read later.

For the convenience of the reader, the three works will be compared and corresponding passages listed in a cumulative index at the end of the second volume of this book, *Mathematical Discovery.*

7. To publish the first part of a book when the second part is not yet available entails certain risks. (There is a German proverb: " Don't show a half-built house to a fool.") These risks are not negligible; yet, in the interest of the practical aim of this work, I decided not to delay the publication of this volume; see p. 210.

This first volume contains Part One of the work, Patterns, and two chapters of Part Two, Toward a General Method.

The four chapters of Part One have more extensive collections of problems than the later chapters. In fact, Part One is in many ways similar to a collection of problems in analysis by G. Szegö and the author (see the Bibliography). There are, however, obvious differences: in the present volume the problems proposed are much more elementary, and methodical points are not only suggested but explicitly formulated and discussed.

The second chapter of Part Two is inspired by a recent work of Werner Hartkopf (see the Bibliography). I present here only some points of Hartkopf's work which seem to me the most engaging, and I present them as they best fit my conception of heuristics, with suitable examples and additional remarks.

8. The Committee on the Undergraduate Program in Mathematics supported the preparation of the manuscript of this book by funds granted by the Ford Foundation. I wish to express my thanks, and I wish to thank the Committee also for its moral support. I wish to thank the editor of the *Journal of Education of the Faculty and College of Education, Vancouver and Victoria*, for permission to incorporate parts of an article into the present work. I also wish to thank Professor Gerald Alexanderson, Santa Clara, California, and Professor Alfred Aeppli, Zurich, Switzerland, for their efficient help in correcting the proofs.

GEORGE POLYA

Zurich, Switzerland
December 1961

HINTS TO THE READER

Section 5 of chapter 2 is quoted as sect. 2.5, subsection (3) of section 5 of chapter 2 as sect. 2.5(3), example 61 of chapter 3 as ex. 3.61.

HSI and MPR are abbreviations for titles of books by the author which will be frequently quoted; see the Bibliography.

Iff. The abbreviation " iff " stands for the phrase " if and only if."

†. The sign † is prefixed to examples, comments, sections, or shorter passages that require more than elementary mathematical knowledge (see the next paragraph). This sign, however, is not used when such a passage is very short.

Most of the material in this book requires only *elementary mathematical knowledge*, that is, as much geometry, algebra, " graphing " (use of co-ordinates), and (sometimes) trigonometry as is (or ought to be) taught in a good high school.

The problems proposed in this book seldom require knowledge beyond the high school level, but, with respect to difficulty, they are often a little above the high school level. The solution is fully (although concisely) presented for some problems, only a few steps of the solution are indicated for other problems, and sometimes only the result is given.

Hints that may facilitate the solution are added to some problems (in parentheses). The surrounding problems may provide hints. Especial attention should be paid to the introductory lines prefixed to the examples (or to certain groups of examples) in some chapters.

The reader who has spent serious effort on a problem may benefit from the effort even if he does not succeed in solving the problem. For example, he may look at some part of the solution, try to extract some helpful information, and then put the book aside and try to work out the rest of the solution by himself.

HINTS TO THE READER

The best time to think about methods may be when the reader has finished solving a problem, or reading its solution, or reading a case history. With his task accomplished and his experience still fresh in mind, the reader, in *looking back* at his effort, can profitably explore the nature of the difficulty he has just overcome. He may ask himself many useful questions: " What was the decisive point? What was the main difficulty? What could I have done better? I failed to see this point: which item of knowledge, which attitude of mind should I have had to see it? Is there some trick worth learning, one that I could use the next time in a similar situation? " All these questions are good, and there are many others— but the best question is the one that comes spontaneously to mind.

CONTENTS

PART TWO
TOWARD A GENERAL METHOD, 115

PART ONE

PATTERNS

*Each problem that I solved became a rule
which served afterwards to solve other problems.*
DESCARTES: *Œuvres*, vol. VI, pp. 20–21; Discours de la Méthode.

*If I found any new truths in the sciences,
I can say that they all follow from, or depend on,
five or six principal problems which I succeeded
in solving and which I regard as so many battles
where the fortune of war was on my side.*
DESCARTES: *op. cit.*, p. 67.

CHAPTER 1

THE PATTERN OF
TWO LOCI

1.1. Geometric constructions

Describing or constructing figures with ruler and compasses has a traditional place in the teaching of plane geometry. The simplest constructions of this kind are used by draftsmen, but otherwise the practical importance of geometric constructions is negligible and their theoretical importance not too great. Still, the place of such constructions in the curriculum is well justified: they are most suitable for familiarizing the beginner with geometric figures, and they are eminently appropriate for acquainting him with the ideas of problem solving. It is for this latter reason that we are going to discuss geometric constructions.

As so many other traditions in the teaching of mathematics, geometric constructions go back to Euclid in whose system they play an important role. The very first problem in Euclid's Elements, Proposition One of Book One, proposes "to describe an equilateral triangle on a given finite straight line." In Euclid's system there is a good reason for restricting the problem to the equilateral triangle but, in fact, the solution is just as easy for the following more general problem: *Describe* (or construct) *a triangle being given its three sides.*

Let us devote a moment to analyzing this problem.

In any problem there must be an *unknown*—if everything is known, there is nothing to seek, nothing to do. In our problem the unknown (the thing desired or required, the *quaesitum*) is a geometric figure, a triangle.

Yet in any problem something must be known or *given* (we call the given things the *data*)—if nothing is given, there is nothing by which we could recognize the required thing: we would not know it if we saw it. In our problem the data are three "finite straight lines" or line segments.

3

Finally, in any problem there must be a *condition* which specifies how the unknown is linked to the data.   In our problem, the condition specifies that the three given segments must be the sides of the required triangle.

The condition is an essential part of the problem.   Compare our problem with the following: "Describe a triangle being given its three altitudes." In both problems the data are the same (three line segments) and the unknown is a geometric figure of the same kind (a triangle).   Yet the connection between the unknown and the data is different, the condition is different, and the problems are very different indeed (our problem is easier).

The reader is, of course, familiar with the solution of our problem. Let a, b, and c stand for the lengths of the three given segments.   We lay down the segment a between the endpoints B and C (draw the figure yourself).   We draw two circles, one with center C and radius b, the other with center B and radius c; let A be one of their two points of intersection. Then ABC is the desired triangle.

1.2. From example to pattern

Let us look back at the foregoing solution, and let us look for promising features which have some chance to be useful in solving similar problems.

By laying down the segment a, we have already located two vertices of the required triangle, B and C; just one more vertex remains to be found. In fact, by laying down that segment we have transformed the proposed problem into another problem equivalent to, but different from, the original problem.   In this new problem

the unknown is a point (the third vertex of the required triangle);
the data are two points (B and C) and two lengths (b and c);
the condition requires that the desired point be at the distance b from the given point C and at the distance c from the given point B.

This condition consists of two parts, one concerned with b and C, the other with c and B.   *Keep only one part of the condition, drop the other part; how far is the unknown then determined, how can it vary?*   A point of the plane that has the given distance b from the given point C is neither completely determined nor completely free: it is restricted to a "locus"; it must belong to, but can move along, the periphery of the circle with center C and radius b.   The unknown point must belong to two such loci and is found as their intersection.

We perceive here a pattern (the "pattern of two loci") which we can imitate with some chance of success in solving problems of geometric construction:

First, reduce the problem to the construction of ONE point.

Then, split the condition into TWO parts so that each part yields a locus for the unknown point; each locus must be either a straight line or a circle.

Examples are better than precepts—the mere statement of the pattern cannot do you much good. The pattern will grow in color and interest and value with each example to which you apply it successfully.

1.3. Examples

Almost all the constructions which traditionally belong to the high school curriculum are straightforward applications of the pattern of two loci.

(1) *Circumscribe a circle about a given triangle.* We reduce the problem to the construction of the center of the required circle. In the so reduced problem

the unknown is a point, say X;
the data are three points A, B, and C;
the condition consists in the equality of three distances:

$$XA = XB = XC$$

We split the condition into two parts:

First $\qquad XA = XB$
Second $\qquad XA = XC$

To each part of the condition corresponds a locus. The first locus is the perpendicular bisector of the segment AB, the second that of AC. The desired point X is the intersection of these two straight lines.

We could have split the condition differently: first, $XA = XB$, second, $XB = XC$. This yields a different construction. Yet can the result be different? Why not?

(2) *Inscribe a circle in a given triangle.* We reduce the problem to the construction of the center of the required circle. In the so reduced problem

the unknown is a point, say X;
the data are three (infinite) straight lines a, b, and c;
the condition is that the point X be at the same (perpendicular) distance from all three given lines.

We split the condition into two parts:

First, X is equidistant from a and b.

Second, X is equidistant from a and c.

The locus of the points satisfying the first part of the condition consists of *two* straight lines, perpendicular to each other: the bisectors of the angles included by a and b. The second locus is analogous. The two loci have four points of intersection: besides the center of the inscribed circle of the triangle we obtain also the centers of the three escribed circles.

Observe that this application calls for a slight modification of our formulation of the pattern at the end of sect. 1.2. What modification?

(3) *Given two parallel lines and a point between them. Draw a circle that is tangent to both given lines and passes through the given point.* If we visualize the required figure (it helps to have it on paper) we may observe that we can easily *solve a part of the problem*: the distance of the two given parallels is obviously the diameter of the required circle and half this distance is the radius.

We reduce the problem to finding the center X of the unknown circle. Knowing the radius, say r, we split the condition as follows:

First, X is at the distance r from the given point.

Second, X is at the distance r from both given lines.

The first part of the condition yields a circle, the second part a straight line midway between, and parallel to, the two given parallels.

Without knowing the radius of the desired circle, we could have split up the condition as follows:

First, X is at the same distance from the given point and the first given line.

Second, X is at the same distance from the given point and the second given line.

Splitting the condition into these two parts is logically unobjectionable but nevertheless useless: the corresponding loci are *parabolas*; we cannot draw them with ruler and compasses—it is an essential part of the scheme that the loci obtained should be circular or rectilinear.

This example may contribute to a better understanding of the pattern of two loci. This pattern helps in many cases, but not in all, as appropriate examples show.

1.4. Take the problem as solved

Wishful thinking is imagining good things you don't have. A hungry man who had nothing but a little piece of dry bread said to himself: "If I had some ham, I could make some ham-and-eggs if I had some eggs."

People tell you that wishful thinking is bad. Do not believe it, this is just one of those generally accepted errors. Wishful thinking may be bad as too much salt is bad in the soup and even a little garlic is bad in the chocolate pudding. I mean, wishful thinking may be bad if there is too much of it or in the wrong place, but it is good in itself and may be a great help in life and in problem solving. That poor guy may enjoy his dry bread more and digest it better with a little wishful thinking about eggs and ham. And we are going to consider the following problem (see Fig. 1.1).

Given three points A, B, and C. Draw a line intersecting AC in the point X and BC in the point Y so that

$$AX = XY = YB$$

Imagine that we knew the position of one of the two points X and Y (this is wishful thinking). Then we could easily find the other point (by drawing a perpendicular bisector). The trouble is that we know neither of the two—the problem does not look easy.

Let us indulge in a little more wishful thinking and *take the problem as solved.* That is, assume that Fig. 1.1 is drawn according to the condition laid down by our problem, so that the three segments of the broken line $AXYB$ are exactly equal. Doing so we imagine a good thing we have not got yet: we imagine that we have found the required location of the line XY; in fact, we imagine that we *have found the solution.*

Yet it is good to have Fig. 1.1 before us. It shows all the geometric elements we should examine, the elements we have and the elements we want, the data and the unknown, assembled as specified by the condition. With the figure before us, we can speculate as to which useful elements we could construct from the data, and which elements could be used in constructing the unknown. We can start from the data and work forward, or start from the unknown and work backward—even side trips could be instructive.

Could you put together at least a few pieces of the jigsaw puzzle? *Could you solve some part of the problem?* There is a triangle in Fig. 1.1, $\triangle XCY$. Can we construct it? We would need three data but, unfortunately, we have only one (the angle at C).

Use what you have, you cannot use what you have not. *Could you derive something useful from the data?* Well, it is easy to join the given points A and B, and the connecting line has some chance to be useful; let us draw it (Fig. 1.2). Yet it is not so easy to see *how* the line AB can be useful—should we rather drop it?

Figure 1.1 looks so empty. There is little doubt that more lines will be needed in the desired construction—what lines?

The lines AX, XY, and YB are equal (we regard them as equal—wishful thinking!). Yet they are in such an awkward relative position—equal lines can be arranged to form much nicer figures. Perhaps we should add more equal lines—or just one more equal line to begin with.

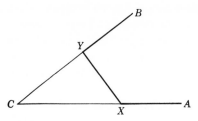

Fig. 1.1. Unknown, data, condition.

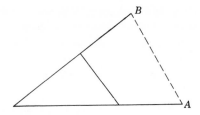

Fig. 1.2. Working forward (from the data).

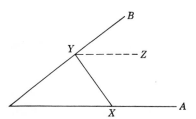

Fig. 1.3. Working backward (from the unknown).

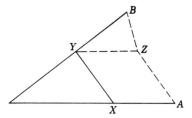

Fig. 1.4. Contacts with previous knowledge.

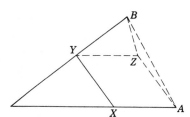

Fig. 1.5. Superposition.

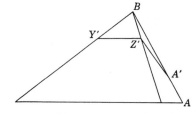

Fig. 1.6. Stepping stone.

Chance or inspiration may prompt us to introduce a line into the picture which, on the face of it, fits quite well into the intended connection: draw *YZ* parallel and equal to *XA*, see Fig. 1.3. (We are starting now from the desired unknown—wishful thinking—and trying to work backward toward the data.)

Introducing the line *YZ* was a trial. Yet the line does not look bad; it brings in familiar shapes. Join *Z* to *A* and *B*, see Fig. 1.4; we obtain the rhombus *XAZY* and the isosceles triangle *BYZ*. *Could you solve some part of the problem?* Can we construct △*BYZ*? We would need two data for an isosceles triangle but, unfortunately, we have only one (the angle at Y is equal to the given angle at C). Still, we have something here. Even if we do not know $\triangle BYZ$ completely, we know its shape; although we do not know its size, we could construct a triangle similar to it. This may bring us a little nearer to the solution, but we have not got it yet: we must try a few more things. Sooner or later we may remember a former trial, the Fig. 1.2. How about combining it with later remarks? By superposing Figs. 1.2 and 1.4 we obtain Fig. 1.5 in which there is a new triangle, $\triangle BZA$. Can we construct it? We could, if we knew $\triangle BYZ$; in that favorable case, we could muster three data: two sides, ZB and $ZA = ZY$, and the angle at B. Well, we do not know $\triangle BZA$; at any rate, we do not know it completely, we know only its shape. Yet then, we can...

We can draw the quadrilateral $BY'Z'A'$, see Fig. 1.6, similar to the quadrilateral $BYZA$ in Fig. 1.5, which is an essential part of the desired configuration. This may be a stepping stone!

1.5. The pattern of similar figures

We carry out the construction, the discovery of which is told by the sequence of Figs. 1.1–1.6.

On the given line BC, see Fig. 1.6, we choose a point Y' at random (but not too far from B). We draw the line $Y'Z'$ parallel to CA so that

$$Y'Z' = Y'B$$

Then, we determine a point A' on AB so that

$$A'Z' = Y'Z'$$

Draw a parallel to $A'Z'$ through A and determine its intersection with the prolongation of the line BZ': this intersection is the desired point Z. The rest is easy.

The two quadrilaterals $AZYB$ and $A'Z'Y'B$ are not only similar but also "similarly located" (homothetic). The point B is their center of similarity. That is, any line connecting corresponding points of the two similar figures has to pass through B.

Here is a remark from which we can learn something about problem solving: Of the two similar figures, the one that came to our attention first, $AZYB$, was actually constructed later.[1]

The foregoing example suggests a general pattern: *If you cannot construct the required figure, think of the possibility of constructing a figure* SIMILAR *to the required figure.*

[1] In this "case history" which we have just finished (we started it in sect. 1.4) the most noteworthy step was to "take the problem as solved." For further remarks on this, cf. HSI, Figures 2, pp. 104–105, and Pappus, pp. 141–148, especially pp. 146–147.

There are examples at the end of this chapter which, if you work them through, may convince you of the usefulness of this pattern of "similar figures."

1.6. Examples

The following examples differ from each other in several respects; their differences may show up more clearly the common feature that we wish to disentangle.

(1) *Draw common tangents to two given circles.* Two circles are given in position (plotted on paper). We wish to draw straight lines touching both circles. If the given circles do not overlap they have four common tangents, two exterior and two interior tangents. Let us confine our attention to the exterior common tangents, see Fig. 1.7, which exist unless one of the two given circles lies completely within the other.

If you cannot solve the proposed problem, look around for an appropriate related problem. There is an obvious related problem (of which the reader is supposed to know the solution): to draw tangents to a given circle from an outside point. This problem is, in fact, a limiting case or *extreme case* of the proposed problem: one of the two given circles is shrunken into a point. We arrive at this extreme case in the most natural way by *variation of the data.* Now we can vary the data in many ways: decrease one radius and leave the other unchanged, or decrease one radius and increase the other, or decrease both. And so we may hit upon the idea of letting both radii decrease *at the same rate,* uniformly, so that both are diminished by the same length in the same time. Visualizing this change, we may observe that each common tangent is shifting, but remains parallel to itself while shifting, till ultimately Fig. 1.8 appears—and here is the solution: draw tangents from the center of the smaller given circle to a new circle which is concentric with the larger given circle and the radius of which is

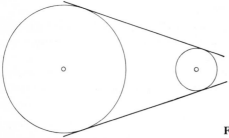

Fig. 1.7. Unknown, data, condition.

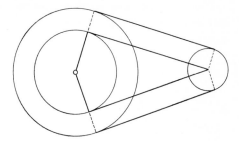

Fig. 1.8. Stepping stone.

the difference of the given radii. Use the figure so obtained as a stepping stone: the step from it to the desired figure is easy (there are just two rectangles to construct).

(2) *Construct a triangle being given the three medians.* We "take the problem as solved"; that is, we draw the (desired) triangle in which the three (given) medians are duly assembled; see Fig. 1.9. We should recollect that the three medians meet in one point (the point M in Fig. 1.9, the centroid of the triangle) which divides each median in the proportion 1:2. To visualize this essential fact, let us mark the midpoint D of the segment AM; the points D and M divide the median AE into three equal parts; see Fig. 1.10.

The desired triangle is divided into six small triangles. *Could you solve a part of the problem?* To construct one of those small triangles we need three data; in fact, we know two sides: one side is one third of a given median, another side is two thirds of another given median—but we do not see a third known piece. Could we introduce some other triangle with three known data? There is the point D in Fig. 1.10 which is obviously eager for more connections—if we join it to a neighboring point we may notice $\triangle MDG$ each side of which is one third of a median—and so we can construct it, from three known sides—here is a stepping stone! The rest is easy.

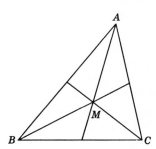

Fig. 1.9. Unknown, data, condition.

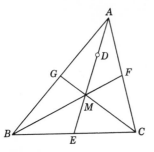

Fig. 1.10. A point eager for more connections.

(3) To each problem concerned with ordinary triangles there corresponds a problem concerned with spherical triangles or trihedral angles. (A trihedral angle is contained between three planes; a sphere described about its vertex as center intersects it in a spherical triangle.) These problems of solid geometry may be reduced to problems of plane geometry. Such reduction of problems about figures in space to drawings in a plane is, in fact, the object of *descriptive geometry*, which is an interesting branch of geometry indispensable to engineers and architects for the accurate drafting of machinery, vessels, buildings, and so on.

The reader needs no knowledge of descriptive geometry, just a little solid geometry and some common sense, to solve the following problem: *Being given the three face angles of a trihedral angle, construct its dihedral angles.*

Let a, b, and c denote the face angles of the trihedral angle (the sides of the corresponding spherical triangle) and α the dihedral angle opposite to the face a (α is an angle of the spherical triangle). Being given a, b, c, construct α. (The same method can serve to construct all three dihedral angles, and so we restrict ourselves to one of them, to α.)

To visualize the *data*, we juxtapose the three angles b, a, and c in a plane; see Fig. 1.11. To visualize the *unknown*, we should see the configuration in space. (Reproduce Fig. 1.11 on cardboard, crease the line between a and b and also that between a and c, and then fold the cardboard to form the trihedral angle.) In Fig. 1.12, the trihedral angle is seen in perspective; A is a point chosen at random on the edge opposite the face a; two perpendiculars to this edge starting from A, one drawn in the face b, the other drawn in the face c, include the angle α that we are required to construct.

Look at the unknown! —It is an angle, the angle α in Fig. 1.12.

What can you do to get this kind of unknown? —We often determine an angle from a triangle.

Is there a triangle in the figure? —No, but we can introduce one.

In fact, there is an obvious way to introduce a triangle: the plane that

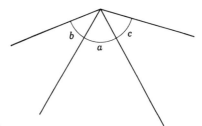

Fig. 1.11. The data.

contains the angle α intersects the trihedral angle in a triangle; see Fig. 1.13. This triangle is a promising auxiliary figure, a likely stepping stone.

In fact, the solution is not far. Return to the figure in the plane, to Fig. 1.11, where the data, the angles a, b, and c, appear in true magnitude. (Unfold the cardboard model we have folded together in passing from Fig. 1.11 to Fig. 1.12.) The point A appears twice, as A_1 and A_2 (by unfolding, we have separated the two faces b and c which are adjacent in space). These points A_1 and A_2 are at the same distance from the vertex V. A perpendicular to A_1V through A_1 meets the other side of the angle b in C, and B is analogously obtained; see Fig. 1.14. Now we know A_2B, BC, and CA_1, the three sides of the auxiliary triangle introduced in Fig. 1.13, and so we can readily construct it (in dotted lines in Fig. 1.14): it contains the desired angle α.

The problem just discussed is analogous to, and uses the construction of, the simplest problem about ordinary triangles which we discussed in sect. 1.1. We can see herein a sort of justice and a hint about the use of analogy.

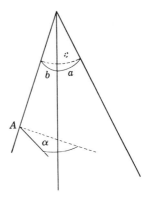

Fig. 1.12. The unknown.

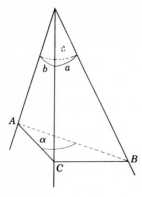

Fig. 1.13. A likely stepping stone.

1.7. The pattern of auxiliary figures

Let us look back at the problems discussed in the foregoing sect. 1.6. They were quite different, and their solutions were quite different too, except that in each case the key to the solution was an *auxiliary figure*: a circle with two tangents from an outside point in (1), a smaller triangle carved out from the desired triangle in (2), another triangle in (3). In each case we could easily construct the auxiliary figure from the data and, once in possession of the auxiliary figure, we could easily construct the originally required figure by using the auxiliary figure. And so we attained our goal in two steps; the auxiliary figure served as a kind of stepping stone; its discovery was the decisive performance, the culminating point of our work. There is a pattern here, the *pattern of auxiliary*

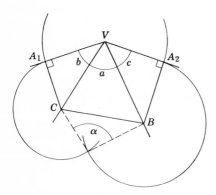

Fig. 1.14. The solution.

figures, which has some promise and which we can describe as follows: *Try to discover some part of the figure or some closely* RELATED FIGURE *which you can construct and which you can use as a stepping stone in constructing the original figure.*

This pattern is very general. In fact, the pattern of similar figures formulated in sect. 1.5 is just a particular case: a figure similar to the required figure is related to it in a particular manner and can serve as a particularly handy auxiliary figure.

Unavoidably, its greater generality renders the pattern of auxiliary figures less concrete, less tangible: it gives no specific advice about what kind of figure we should seek. Experience, of course, can give us some directives (although no hard and fast rules): we should look for figures which are easy to "carve out" from the desired figure, for "simple" figures (as triangles), for "extreme cases," and so on. We may learn procedures, such as the variation of the data, or the use of analogy, which, in certain cases, may indicate an appropriate auxiliary figure.

We have now isolated three different patterns which we may use in dealing with problems of geometric construction. The pattern of *auxiliary figures* leaves us more choice, but offers a less definite target, than the pattern of *similar figures*. The pattern of *two loci* is the simplest—you may try it first just because, in most cases, it is best to try the simplest thing first. Yet do not commit yourself, keep an open mind: take the problem as solved, draw a figure in which the *unknown* and the *data* are appropriately assembled, each element at its right place, all elements connected by the right relations, as required by the *condition*. Study this figure, try to recognize in it some familiar configuration, try to recall any relevant knowledge you may have (related problems, applicable theorems), look out for an opening (a more accessible part of the figure, for instance). You may be lucky: a bright idea may emerge from the figure and suggest an appropriate auxiliary line, the suitable pattern, or some other useful step.

Examples and Comments on Chapter 1

1.1. What is the locus of a variable point that has a given distance from a given point?

1.2. What is the locus of a variable point that has a given distance from a given straight line?

1.3. A variable point remains equidistant from two given points; what is its locus?

1.4. A variable point remains equidistant from two given parallel straight lines; what is its locus?

1.5. A variable point remains equidistant from two given intersecting straight lines; what is its locus?

1.6. Of a triangle, given two vertices, A and B, and the angle γ, opposite to the side AB; the triangle is not determined, its third vertex (that of γ) can vary. What is the locus of this third vertex?

1.7. *Notation.* In dealing with a triangle, it is convenient to use the following notation:

$A,$	$B,$	C	vertices
$a,$	$b,$	c	sides
$\alpha,$	$\beta,$	γ	angles
$h_a,$	$h_b,$	h_c	altitudes ("heights")
$m_a,$	$m_b,$	m_c	medians
$d_\alpha,$	$d_\beta,$	d_γ	bisectors of the angles ("disectors"?!)
		R	radius of circumscribed circle
		r	radius of inscribed circle

It is understood that the side a is opposite the angle α, the vertex of which is the point A which is the common endpoint of the three lines h_a, m_a, and d_α. According to common usage, a stands both for the side (a line segment) and for the length of the side; the reader has to find out from the context which meaning is intended. The same ambiguity is inherent in the symbols $b, c, h_a, \ldots$ d_γ, R, r. We follow this usage although it is objectionable.

The problem "Triangle from a, b, c" means, of course, "construct a triangle being given $a, b,$ and c." Observe that there may be no solution (the figure satisfying the proposed condition may not exist) if the data are adversely chosen; for example, there is no triangle with given sides $a, b,$ and c, if $a > b + c$. Experiment first with data for which the required figure is likely to exist.

1.8. Triangle from a, b, m_a.

1.9. Triangle from a, h_a, m_a.

1.10. Triangle from a, h_a, α.

1.11. Triangle from a, m_a, α.

1.12. Given three (infinite) straight lines. Construct a circle that touches the first two lines and has its center on the third line.

1.13. Given two intersecting infinite straight lines and a line segment of length r. Construct a circle with radius r that touches the two given lines.

1.14. Construct a circle, being given one point on it, one straight line tangent to it, and its radius.

1.15. *Three lighthouses* are visible from a ship; their positions on the map are known, and the angles between the rays of light coming from them have been measured. Plot the position of the ship on the map.

1.16. Within a given circle, describe three equal circles so that each shall touch the other two and also the given circle. (This figure can sometimes be seen in Gothic tracery where analogous figures, with four or six inner circles, are more frequent.)

1.17. Inside a given triangle find a point from which all three sides are seen under the same angle.

1.18. Trisect the area of a given triangle.

That is, you should locate a point X inside the given $\triangle ABC$ so that $\triangle XBC$, $\triangle XCA$, and $\triangle XAB$ are equal in area.

[*Keep only a part of the condition, drop the other part:* if only the two triangles $\triangle XCA$ and $\triangle XCB$ are supposed to be equal, what is the locus of X? The answer to this question may show you a way to the solution, but there are also other approaches.]

1.19. Triangle from a, α, r.

[*Keep only a part of the condition, drop the other part:* disregard r, but keep a and α; what is the locus of the center of the inscribed circle?]

1.20. Triangle from a, h_b, c.

1.21. Triangle from a, h_b, d_γ.

1.22. Triangle from a, h_b, h_c.

1.23. Triangle from h_a, h_b, β.

1.24. Triangle from h_a, β, γ.

1.25. Triangle from h_a, d_α, α.

1.26. Construct a parallelogram, being given one side and two diagonals.

1.27. Construct a trapezoid being given its four sides a, b, c, and d; a and c should be parallel.

1.28. Construct a quadrilateral being given a, b, c, and d, its four sides, and the angle ϵ, included by the opposite sides a and c produced.

1.29. Triangle from a, $b + c$, α.

[Do not fail to introduce *all the data* into the figure. Where is the "right place" for $b + c$?]

1.30. Triangle from a, $b + c$, $\beta - \gamma$.

1.31. Triangle from $a + b + c$, h_a, α.

[Symmetry: b and c (not given) play interchangeable roles.]

1.32. Given two circles exterior to each other, draw their interior common tangents. (The two circles are situated in the same halfplane with respect to an exterior common tangent, in different halfplanes with respect to an interior common tangent.)

1.33. Given three equal circles, construct a circle containing, and tangent to, all three given circles.

1.34. Triangle from α, β, d_γ.

1.35. Inscribe a square in a given right triangle.   One corner of the square is required to coincide with the right-angle corner of the given triangle, the opposite vertex of the square should lie on the hypotenuse, the two other vertices on the legs of the right triangle, one on each.

1.36. Inscribe a square in a given triangle ABC.   Two vertices of the square are required to lie on AB, one on AC, and one on BC.

1.37. Inscribe a square in a given sector of a circle.   Two vertices of the square are required to lie on the arc, one vertex on each of the two sides of the central angle of the sector.

1.38. Construct a circle being given two points on it and one straight line tangent to it.

1.39. Construct a circle, being given one point on it and two straight lines tangent to it.

1.40. Construct a pentagon circumscribable about a circle being given its five angles α, β, γ, δ, and ϵ (subject, of course, to the condition $\alpha + \beta + \gamma + \delta + \epsilon = 540°$) and the length of its perimeter l.

1.41. Triangle from h_a, h_b, h_c.

1.42. *A flaw.*   It may happen that a problem of geometric construction has no solution: there may be no figure satisfying the proposed condition with the proposed data.   For instance, there exists no triangle the sides of which have the given lengths a, b, and c if $a > b + c$.   A perfect method of solution will either obtain a figure satisfying the proposed condition or show in failing that there exists no such figure.

There can arise, however, the following situation: the proposed problem itself does possess a solution, yet an auxiliary problem does not—an auxiliary figure, which our scheme would need for the construction of the originally required figure, is impossible to construct.   This is, of course, a flaw in our scheme.

Is your method for solving ex. 1.41 perfect in this respect?   (The triangle with sides 65, 156, 169 is a right triangle—the sides are proportional to 5, 12, 13—with heights 156, 65, 60.)   If the answer is No, can you improve your method?

1.43. Triangle from a, α, R.

1.44. *Looking back* at the solution of ex. 1.43, you may ask some instructive questions and propose some related problems.

(*a*) An analogous problem?

(*b*) A more general problem?

(*c*) Triangle from a, β, R.

(*d*) Triangle from a, r, R.

1.45. *Three listening posts.* The time at which the sound of an enemy gun is heard has been exactly observed at three listening posts A, B, and C. On the basis of these data, plot on the map the position X of that enemy gun.

Regard the velocity of sound as known. Explain the analogy with, and the difference from, the problem of the three lighthouses, ex. 1.15.

1.46. *On the pattern of two loci.* Are the loci with which exs. 1.2, 1.5, and 1.6 are concerned usable in connection with the pattern of two loci? Cf. the statement at the end of sect. 1.2.

1.47. *The pattern of three loci.* A concept of plane geometry may have various analogues in solid geometry. For instance, in sect. 1.6(3) we regarded a spherical triangle or a trihedral angle as analogous to an ordinary plane triangle. Yet we could also regard a tetrahedron as analogous to an ordinary triangle; seen from this viewpoint, the following problem appears as analogous to the problem of sect. 1.3(1).

Circumscribe a sphere about a given tetrahedron.

Let us work out the analogy in some detail. We reduce the problem to obtaining the center of the required sphere. In the so reduced problem

the unknown is a point, say X;

the data are four points (the vertices of the given tetrahedron), say A, B, C, and D;

the condition consists in the equality of four distances

$$XA = XB = XC = XD$$

We may split this condition into three parts:

First	$XA = XB$
Second	$XA = XC$
Third	$XA = XD$

To each part of the condition corresponds a locus. If the point X satisfies the first part of the condition, its locus is (it can vary on) a plane, the perpendicular bisector of the segment AB; to each other part of the condition there corresponds an analogous plane. Finally, the desired center of the sphere is obtained as the intersection of three planes.

Let us assume that we have instruments with which we can determine the points of intersection of three given surfaces when each of these surfaces is either a plane or a sphere. (In fact, we have made this assumption implicitly in the foregoing. By the way, ruler and compasses are such instruments—we can determine with them those points of intersection if we know enough descriptive geometry.) Then we can propose and solve problems of geometric construction in space. The foregoing problem is an example and its solution sets an example: with the help of analogy, we can disentangle from it a pattern for solving problems of construction in space, the *pattern of three loci*.

1.48. In the foregoing ex. 1.47, as in the example of sect. 1.3(1), we could have split the condition differently and so obtain another (although a pretty similar) construction. Yet can the result be different? Why not?

1.49. *On geometric constructions.* There are many problems of geometric construction where the required figure obviously "exists" but cannot be constructed with ruler and compasses (it could be constructed with other—equally idealized—instruments). A famous problem of this kind is the trisection of the angle: a *general* angle cannot be divided into three equal parts by ruler and compasses; see Courant and Robbins, pp. 137–138.

A perfect method for geometric constructions should *either* lead us to a construction of the required figure by ruler and compasses *or* show that such a construction is impossible. Our patterns (two loci, similar figures, auxiliary figures) are not useless (as, I hope, the reader has had opportunity to convince himself) but they yield no perfect method; they frequently suggest a construction but, when they do not suggest one, we are left in the dark about the alternative with which we are most concerned: is the construction impossible in itself, or is it possible and just our effort insufficient?

There is a well known more perfect method for geometric constructions (reduction to algebra—but we need not enter upon details now). Yet for another kind of problem which we may face another day there may be no perfect method known at that time—and still we have to try. And so the patterns considered may contribute to the education of the problem solver just by their inherent imperfection.

1.50. *More problems.* Devise some problems similar to, but different from, the problems proposed in this chapter—especially such problems as you can solve.

1.51. *Sets.* We cannot define the concept of a *set* in terms of more fundamental concepts, because there are no more fundamental concepts. Yet, in fact, everybody is familiar with this concept, even if he does not use the word "set" for it. "Set of elements" means essentially the same as "class of objects" or "collection of things" or "aggregate of individuals." "Those students who will make an A in this course" form a set even if, at this moment, you could not tell all their names. "Those points in space that are equidistant from two given points" form a very clearly defined set of points, a plane. "Those straight lines in a given plane that have a given distance from a given point" form an interesting set consisting of all the tangents of a certain circle. If *a*, *b*, and *c* are any three distinct objects, the set to which just these three objects belong as elements is clearly defined.

Two sets are *equal* if every object that belongs to one of them belongs also to the other. If any element that belongs to the set *A* belongs also to the set *B*, we say that *A* is *contained* in *B*; there are many ways to say the same thing: *B* contains *A*, *B* includes *A*, *A* is a *subset* of *B*, and so on.

It is often convenient to consider the *empty set*, that is, the set to which no element belongs. For example, the "set of those students who will make A in this course" could well turn out to be the empty set, if no student makes a better grade than B, or if the course should be discontinued without a final examination. The empty set is a useful set as 0 is a useful number. Now, 0 is less than any positive integer; similarly, the empty set is considered as a subset of any set.

The greatest common subset of several sets is termed their *intersection*. That is, the intersection of the sets A, B, C,..., and L consists of those, and only those, elements that belong simultaneously to each of the sets A, B, C,..., and L.

For example, let A and B denote two planes, each considered as a set of points; if they are different and nonparallel, their intersection is a straight line; if they are different but parallel, their intersection is the empty set; if they are identical, their "intersection" is identical with any of them. If A, B, and C are three planes and there is no straight line parallel to all three of them, their intersection is a set containing just one element, a point.

The term "locus" means essentially the same as the term "set": the set (or locus) of those points of a plane that have a given distance from a given point is a circle.

In this example, we define the set (or locus) by stating a *condition* that its elements (points) must satisfy, or a *property* that these elements must possess: the points of a circle satisfy the condition, or have the property, that they are all contained in the same plane and all have the same distance from a given point.

The concepts of "condition" and "property" are indissolubly linked with the concept of a set. In many mathematical examples we can clearly and simply state the condition or property that characterizes the elements of a set. Yet, if a more informative description is lacking we can always say: the elements of the set S have the property of belonging to S, and satisfy the condition that they belong to S.

The consideration of the pattern of three loci (after that of two loci; see ex. 1.47) may have given us already a hint of a wider generalization. The consideration of sets and their intersections intensifies the suggestion. We now leave this suggestion to mature in the mind of the reader and we shall return to it in a later chapter.

(The least extensive set of which each one of several given sets is a subset is called the *union* of those given sets. That is, the union of the sets A, B, ..., and L contains all the elements of A, all the elements of B,..., and all the elements of L, and any element that the union contains, must belong to at least one of the sets A, B,..., and L (it may belong to several of them).

Intersection and union of sets are closely allied concepts (they are "complementary" concepts in a sense which we cannot but hint), and we could not very well discuss one without mentioning the other. In fact, we shall have more opportunity to consider the intersection of given sets than their union. The reader should familiarize himself from some other book with the first notions of the theory of sets which may be introduced into the high schools in the near future.)

CHAPTER 2

THE CARTESIAN PATTERN

2.1. Descartes and the idea of a universal method

René Descartes (1596–1650) was one of the very great. He is regarded by many as the founder of modern philosophy, his work changed the face of mathematics, and he also has a place in the history of physics. We are here mainly concerned with one of his works, the *Rules for the Direction of the Mind* (cf. ex. 2.72).

In his "Rules," Descartes planned to present a universal method for the solution of problems. Here is a rough outline of the scheme that Descartes expected to be applicable to all types of problems:

First, reduce any kind of problem to a mathematical problem.

Second, reduce any kind of mathematical problem to a problem of algebra.

Third, reduce any problem of algebra to the solution of a single equation.

The more you know, the more gaps you can see in this project. Descartes himself must have noticed after a while that there are cases in which his scheme is impracticable; at any rate, he left unfinished his "Rules" and presented only fragments of his project in his later (and better known) work *Discours de la Méthode*.

There seems to be something profoundly right in the intention that underlies the Cartesian scheme. Yet it is more difficult to carry this intention into effect, there are more obstacles and more intricate details than Descartes imagined in his first enthusiasm. Descartes' project failed, but it was a great project and even in its failure it influenced science much more than a thousand and one little projects which happened to succeed.

Although Descartes' scheme does not work in all cases, it does work in an inexhaustible variety of cases, among which there is an inexhaustible variety of *important* cases.   When a high school boy solves a "word problem" by "setting up equations," he follows Descartes' scheme and in doing so he prepares himself for serious applications of the underlying idea.

And so it may be worthwhile to have a look at some high school work.

2.2. A little problem

Here is a brain teaser which may amuse intelligent youngsters today as it probably amused others through several centuries.

A farmer has hens and rabbits.   These animals have 50 heads and 140 feet.   How many hens and how many rabbits has the farmer?

We consider several approaches.

(1) *Groping.*   There are 50 animals altogether.   They cannot all be hens, because then they would have only 100 feet.   They cannot all be rabbits, because they would then have 200 feet.   Yet there should be just 140 feet.   If just one half of the animals were hens and the other half rabbits, they would then have. . . .   Let us survey all these cases in a table:

Hens	Rabbits	Feet
50	0	100
0	50	200
25	25	150

If we take a smaller number of hens, we have to take a larger number of rabbits and this leads to more feet.   On the contrary, if we take a larger number of hens. . . .   Yes, there must be more than 25 hens—let us try 30:

Hens	Rabbits	Feet
30	20	140

I have got it!   Here is the solution!

Yes, indeed, we have got the solution, because the given numbers, 50 and 140, are relatively small and simple.   Yet if the problem, proposed with the same wording, had larger or more complicated numbers, we would need more trials or more luck to solve it in this manner, by merely muddling through.

(2) *Bright idea.*   Of course, our little problem can be solved less "empirically" and more "deductively"—I mean with fewer trials, less guesswork, and more reasoning.   Here is another solution.

The farmer surprises his animals in an extraordinary performance: each hen is standing on one leg and each rabbit is standing on its hind legs. In this remarkable situation just one half of the legs are used, that is, 70

legs. In this number 70 the head of a hen is counted just *once* but the head of a rabbit is counted *twice*. Take away from 70 the number of all heads, which is 50; there remains the number of the rabbit heads—there are

$$70 - 50 = 20$$

rabbits! And, of course, 30 hens.

This solution would work just as well if the numbers in our little problem (50 and 140) were replaced by less simple numbers. This solution (which can be presented less whimsically) is ingenious: it needs a clear intuitive grasp of the situation, a little bit of a bright idea—my congratulations to a youngster of fourteen who discovers it by himself. Yet bright ideas are rare—we need a lot of luck to conceive one.

(3) *By algebra.* We can solve our little problem without relying on chance, with less luck and more system, if we know a little algebra.

Algebra is a language which does not consist of words but of symbols. If we are familiar with it we can translate into it appropriate sentences of everyday language. Well, let us try to translate into it the proposed problem. In doing so, we follow a precept of the Cartesian scheme: "reduce any kind of problem to a problem of algebra." In our case the translation is easy.

State the problem

in English	*in algebraic language*
A farmer has	
a certain number of hens	x
and a certain number of rabbits	y
These animals have fifty heads	$x + y = 50$
and one hundred forty feet	$2x + 4y = 140$

We have translated the proposed question into a system of two equations with two unknowns, x and y. Very little knowledge of algebra is needed to solve this system: we rewrite it in the form

$$x + 2y = 70$$
$$x + \ y = 50$$

and subtracting the second equation from the first we obtain

$$y = 20$$

Using this we find, from the second equation of the system, that

$$x = 30$$

This solution works just as well for large given numbers as for small ones, works for an inexhaustible variety of problems, and needs no rare bright idea, just a little facility in the use of the algebraic language.

(4) *Generalization.* We have repeatedly considered the possibility of substituting other, especially larger, numbers for the given numbers of our problem, and this consideration was instructive. It is even more instructive to substitute *letters for the given numbers.*

Substitute *h* for 50 and *f* for 140 in our problem. That is, let *h* stand for the number of heads, and *f* for the number of feet, of the farmer's animals. By this substitution, our problem acquires a new look; let us consider also the translation into algebraic language.

A farmer has	
a certain number of hens	x
and a certain number of rabbits.	y
These animals have *h* heads	$x + y = h$
and *f* feet.	$2x + 4y = f$

The system of two equations that we have obtained can be rewritten in the form

$$x + 2y = \frac{f}{2}$$
$$x + y = h$$

and yields, by subtraction,

$$y = \frac{f}{2} - h$$

Let us retranslate this formula into ordinary language: the number of rabbits equals one half of the number of feet, less the number of heads: this is the result of the imaginative solution (2).

Yet here we did not need any extraordinary stroke of luck or whimsical imagination; we attained the result by a straightforward routine procedure after a simple initial step which consisted in replacing the given numbers by letters. This step is certainly simple, but it is an important step of generalization.[1]

(5) *Comparison.* It may be instructive to compare different approaches to the same problem. Looking back at the four preceding approaches, we may observe that each of them, even the very first, has some merit, some specific interest.

[1] Cf. HSI, Generalization 3, pp. 109–110; Variation of the problem 4, pp. 210–211; Can you check the result? 2, p. 60.

The first procedure which we have characterized as "groping" and "muddling through" is usually described as a solution by *trial and error*. In fact, it consists of a series of trials, each of which attempts to correct the error committed by the preceding and, on the whole, the errors diminish as we proceed and the successive trials come closer and closer to the desired final result. Looking at this aspect of the procedure, we may wish a better characterization than "trial and error"; we may speak of "successive trials" or "successive corrections" or "successive approximations." The last expression may appear, for various reasons, to be the most suitable. The term *method of successive approximations* naturally applies to a vast variety of procedures on all levels. You use successive approximations when, in looking for a word in the dictionary, you turn the leaves and proceed forward or backward according as a word you notice precedes or follows in alphabetical order the word you are looking for. A mathematician may apply the term successive approximations to a highly sophisticated procedure with which he tries to treat some very advanced problem of great practical importance that he cannot treat otherwise. The term even applies to science as a whole; the scientific theories which succeed each other, each claiming a better explanation of phenomena than the foregoing, may appear as successive approximations to the truth.

Therefore, the teacher should not discourage his students from using "trial and error"—on the contrary, he should encourage the intelligent use of the fundamental method of successive approximations. Yet he should convincingly show that for such simple problems as that of the hens and rabbits, and in many more (and more important) situations, straightforward algebra is more efficient than successive approximations.

2.3. Setting up equations

In the foregoing, cf. sect. 2.2(3), we have translated a proposed problem from the ordinary language of words into the algebraic language of symbols. In our example, the translation was obvious; there are cases, however, where the translation of the problem into a system of equations demands more experience, or more ingenuity, or more work.[2]

What is the nature of this work? Descartes intended to answer this question in the second part of his "Rules" which, however, he left unfinished. I wish to extract from his text and present in contemporary language such parts of his considerations as are the most relevant at this stage of our study. I shall leave aside many things that Descartes did say, and I shall make explicit a few things that he did not quite say, but I still think that I shall not distort his intentions.

[2] Cf. HSI, Setting up equations, pp. 174–177.

I wish to follow Descartes' manner of exposition: I shall begin each explanation by a concise "advice" (in fact, it is rather a summary) and then expand that advice (summary) by adding comments.

(1) *First, having well understood the problem, reduce it to the determination of certain unknown quantities* (Rules XIII–XVI).

To spend time on a problem that we do not understand would be foolish. Therefore, our first and most obvious duty is to understand the problem, its meaning, its purpose.

Having understood the problem as a whole, we turn our attention to its principal parts.   We should see very clearly

what kind of thing we have to find (the UNKNOWN or unknowns)

what is given or known (the DATA)

how, by what relations, the unknowns and the data are connected with each other (the CONDITION).

(In the problem of sect. 2.2(4) the unknowns are x and y, and the data h and f, the numbers of hens and rabbits, heads and feet, respectively. The condition is expressed first in words, then in equations.)

Following Descartes, we now confine ourselves to problems in which the unknowns are quantities (that is, numbers but not necessarily integers). Problems of other kinds, such as geometrical or physical problems, may be reduced sometimes to problems of this purely quantitative type, as we shall illustrate later by examples; cf. sects. 2.5 and 2.6.

(2) *Survey the problem in the most natural way, taking it as solved and visualizing in suitable order all the relations that must hold between the unknowns and the data according to the condition* (Rule XVII).

We imagine that the unknown quantities have values fully satisfying the condition of the problem: this is meant essentially by "taking the problem as solved"; cf. sect. 1.4.   Accordingly, we treat unknown and given quantities equally in some respects; we visualize them connected by relations as the condition requires.   We should survey and study these relations in the spirit in which we survey and study the figure when planning a geometric construction; see the end of sect. 1.7.   The aim is to find some indication about our next task.

(3) *Detach a part of the condition according to which you can express the same quantity in two different ways and so obtain an equation between the unknowns.   Eventually you should split the condition into as many parts, and so obtain a system of as many equations, as there are unknowns* (Rule XIX).

The foregoing is a free rendering, or *paraphrase*, of the statement of Descartes' Rule XIX.   After this statement there is a great gap in

Descartes' manuscript: the explanation which should have followed the statement of the Rule is missing (it was probably never written). Therefore, we have to make up our own comments.

The aim is stated clearly enough: we should obtain a system of n equations with n unknowns. It is understood that the computation of these unknowns should solve the proposed problem. Therefore, the system of equations should be equivalent to the proposed condition. If the whole system expresses the whole condition, each single equation of the system should express some part of the condition. Hence, in order to set up the n equations we should split the condition into n parts. But how?

The foregoing considerations under (1) and (2) (which outline very sketchily Descartes' Rules XIII–XVII) give some indications, but no definite instructions. Certainly, we have to understand the problem, we have to see the unknowns, the data, and the condition very, very clearly. We may profit from surveying the various clauses of the condition and from visualizing the relations between the unknowns and the data. All these activities give us a chance to obtain the desired system of equations, but no certainty.

The advice that we are considering (the paraphrase of Rule XIX) stresses an additional point: in order to obtain an equation we have to *express the same quantity in two different ways.* (In the example of sect. 2.2(3) an equation expresses the *number of feet* in two different ways.) This remark, properly digested, often helps to discover an equation between the unknowns—it can always help to explain the equation after it has been discovered.

In short, there are some good suggestions, but there is no foolproof precept for setting up equations. Yet, where no precept helps, practice may help.

(4) *Reduce the system of equations to one equation* (Rule XXI).

The statement of Descartes' Rule XXI which is here paraphrased is not followed by an explanation (in fact, it is the last sentence in Descartes' manuscript). We shall not examine here under which conditions a system of algebraic equations can be reduced to a single equation or how such reduction can be performed; these questions belong to a purely mathematical theory which is more intricate than Descartes' short advice may lead us to suppose, but is pretty well explored nowadays and no concern of ours at this point. Very little algebra will be sufficient to perform the reduction in those simple cases in which we shall need it.

There are other questions which remain unexplored although we should concern ourselves with them. Yet we may take them up more profitably after some examples.

2.4. Classroom examples

The "word problems" of the high school are trivial for mathematicians, but not so trivial for high school boys or girls or teachers. I think, however, that a teacher who makes an earnest effort to bring Descartes' advice, presented in the foregoing, down to classroom level and to put it into practice will avoid many of the usual pitfalls and difficulties.

First of all, the student should not start doing a problem before he has understood it. It can be checked to a certain extent whether the student has really understood the problem: he should be able to repeat the statement of the problem, point out the unknowns and the data, and explain the condition in his own words. If he can do all this reasonably well, he may proceed to the main business.

An equation expresses a *part of the condition*. The student should be able to tell which part of the condition is expressed by an equation that he brings forward—and which part is not yet expressed.

An equation expresses the *same quantity in two different ways*. The student should be able to tell which quantity is so expressed.

Of course, the student should possess the *relevant knowledge* without which he could not understand the problem. Many of the usual high school problems are "rate problems" (see the next three examples). Before he is called upon to do such a problem, the student should acquire in some form the idea of "rate," proportionality, uniform change.

(1) *One pipe can fill a tank in* 15 *minutes, another pipe can fill it in* 20 *minutes, a third pipe in* 30 *minutes. With all three pipes open, how long will it take to fill the empty tank?*

Let us assume that the tank contains g gallons of water when it is full. Then the rate of flow through the first pipe is

$$\frac{g}{15}$$

gallons per minute.   Since

$$\text{amount} = \text{rate} \times \text{time}$$

the amount of water flowing through the first pipe in t minutes is

$$\frac{g}{15} t$$

If the three pipes together fill the empty tank in t minutes, the *amount of water in the full tank* can be expressed in two ways:

$$\frac{g}{15} t + \frac{g}{20} t + \frac{g}{30} t = g$$

The left-hand side shows the contribution of each pipe separately, the right-hand side the joint result of these three contributions.

Division by g yields the equation for the required time t:

$$\frac{t}{15} + \frac{t}{20} + \frac{t}{30} = 1$$

Of course, the derivation of the equation could be presented differently and the problem itself could be generalized and modified in various ways.

(2) *Tom can do a job in* 3 *hours, Dick in* 4 *hours, and Harry in* 6 *hours. If they do it together* (and do not delay each other), *how long does the job take?*

Tom can do $\frac{1}{3}$ of the whole job in one hour; we can also say that Tom is working at the rate of $\frac{1}{3}$ of the job per hour. Therefore, in t hours Tom does $t/3$ of the job. If the three boys work together and finish the work in t hours (and if they do not delay each other—a very iffy condition), the *full amount of work* can be expressed in two ways:

$$\frac{t}{3} + \frac{t}{4} + \frac{t}{6} = 1$$

in fact, the 1 on the right-hand side stands for "one full job."

This problem is almost identical with the foregoing (1), even numerically since

$$15:20:30 = 3:4:6$$

It is instructive to formulate a common generalization of both (using letters). It is also instructive to compare the solutions and weigh the advantage and disadvantage of introducing the quantity g into the solution (1).

(3) *A patrol plane flies* 220 *miles per hour in still air. It carries fuel for* 4 *hours of safe flying. If it takes off on patrol against a wind of* 20 *miles per hour, how far can it fly and return safely?*

It is understood that the wind is supposed to blow with unchanged intensity during the whole flight, that the plane travels in a straight line, that the time needed for changing direction at the furthest point is negligible, and so on. All word problems contain such unstated simplifying assumptions and demand from the problem solver some preliminary work of *interpretation* and *abstraction*. This is an essential feature of the word problems which is not always trivial and should be brought into the open, at least now and then.

The problem becomes more instructive if for the numbers

$$220 \quad 20 \quad 4$$

we substitute general quantities

$$v \quad\quad w \quad\quad T$$

which denote the velocity of the plane in still air, the velocity of the wind, and the total flying time, respectively; these three quantities are the *data*. Let x stand for the distance flown in one direction, t_1 for the duration of the outgoing flight, t_2 for the duration of the homecoming flight; these three quantities are *unknowns*. It is useful to display some of these quantities in a neat arrangement:

	Going	Returning
Distance	x	x
Time	t_1	t_2
Velocity	$v - w$	$v + w$

(To fill out the last line we need, in fact, some "unsophisticated" knowledge of kinematics.) Now, as we should know,

$$\text{distance} = \text{velocity} \times \text{time}$$

We express each of the following three quantities in two ways:

$$x = (v - w)t_1$$
$$x = (v + w)t_2$$
$$t_1 + t_2 = T$$

We have here a system of three equations for the three unknowns x, t_1, and t_2. In fact, only x was required by the proposed problem; t_1 and t_2 are *auxiliary unknowns* which we have introduced in order to express neatly the whole condition. Eliminating t_1 and t_2, we find

$$\frac{x}{v - w} + \frac{x}{v + w} = T$$

and hence

$$x = \frac{(v^2 - w^2)T}{2v}$$

There is no difficulty in substituting numerical values for the data v, w, and T. It is more interesting to examine the result, and to check it by the *variation of the data*.

If $w = 0$, then $2x = vT$. This is right, obviously: the whole flight is now supposed to take place in still air.

If $w = v$, then $x = 0$. Again obvious: against a headwind with speed v, the plane cannot start at all.

If w increases from the value $w = 0$ to the value $w = v$, the distance x decreases steadily, according to the formula. And so, again, the formula agrees with what we can foresee without any algebra, just by visualizing the situation.

Working with numerical data instead of general data (letters) we would have missed this instructive discussion of the formula and the valuable checks of our result.   By the way, there are still other interesting checks.

(4) *A dealer has two kinds of nuts; one costs* 90 *cents a pound, the other* 60 *cents a pound.   He wishes to make* 50 *pounds of a mixture that will cost* 72 *cents a pound.   How many pounds of each kind should he use?*

This is a typical, rather simple "mixture problem."   Let us say that the dealer uses x pounds of nuts of the first kind, and y pounds of the second kind; x and y are the unknowns.   We can conveniently survey the unknowns and the data in the array:

	First kind	Second kind	Mixture
Price per pound	90	60	72
Weight	x	y	50

Express in two ways the *total weight of the mixture*:

$$x + y = 50$$

Then express in two ways the *total price of the mixture*:

$$90x + 60y = 72 \cdot 50$$

We have here a system of two equations for the two unknowns x and y. We leave the solution to the reader, who should have no trouble in finding the values

$$x = 20, \qquad y = 30$$

In passing from "numbers" to "letters" the reader obtains a problem which, as it will turn out later, has still other (and more interesting) interpretations.

2.5. Examples from geometry

We shall discuss just two examples.

(1) *A problem of geometric construction.*   It is possible to reduce any problem of geometric construction to a problem of algebra.   We cannot treat here the general theory of such reduction,[3] but here is an example.

A triangular area is enclosed by a straight line AB and two circular arcs, AC and BC.   The center of one circle is A, that of the other is B, and each circle passes through the center of the other.   Inscribe into this triangular area a circle touching all three boundary lines.

The desired configuration, Fig. 2.1, is sometimes seen in Gothic tracery.

Obviously, we can reduce the problem to the construction of one point: the center of the required circle.   One locus for this point is also obvious:

---

[3] See Courant-Robbins, pp. 117–140.

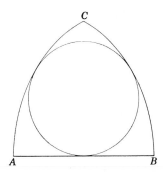

Fig. 2.1. From a Gothic window.

the perpendicular bisector of the segment AB which is a line of symmetry for the given triangular area.   And so there remains to find another locus.

Keep only a part of the condition, drop the other part. We consider a (variable) circle touching not three, but only two boundary lines: the straight line AB and the circular arc BC; see Fig. 2.2.   In order to find the locus of the center of this variable circle, we use analytic geometry.   We let the origin of our rectangular coordinate system coincide with the point A, and let the x axis pass through the point B; see Fig. 2.2.   Let x and y denote the coordinates of the center of the variable circle.   Join this center to the two essential points of contact, one with the straight line AB, the other with the circular arc BC; see Fig. 2.2.   The two radii have the same length which, therefore, can be expressed in two different manners (set $AB = a$):

$$y = a - \sqrt{x^2 + y^2}$$

By getting rid of the square root, we transform this equation into

$$x^2 = a^2 - 2ay$$

And so the locus of the center of the variable circle turns out to be a parabola—a locus of no immediate use in geometric constructions.

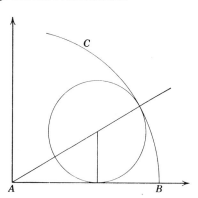

Fig. 2.2. We have dropped a part of the condition.

Yet the obvious locus mentioned at the beginning, the perpendicular bisector of AB, has the equation

$$x = \frac{a}{2}$$

which, combined with the equation of the parabola, yields the ordinate of the desired center of the circle.

$$y = \frac{3a}{8}$$

and this ordinate is easy to construct from the given length $a = AB$.

(2) *The analogue of Pythagoras' theorem in solid geometry.*   Analogy is not unambiguous.   There are various facts of solid geometry which can be quite properly regarded as analogous to the Pythagorean proposition. We arrive at such a fact if we regard a cube as analogous to a square, and a tetrahedron that we obtain by cutting off a corner of the cube by an oblique plane as analogous to a right triangle (which we obtain by cutting off a corner of a square by an oblique straight line).   To the rectangular vertex of the right triangle there corresponds a vertex of the tetrahedron which we shall call a *trirectangular* vertex.   In fact, the three edges of the tetrahedron starting from this vertex are perpendicular to each other, forming three right angles.

Pythagoras' theorem solves the following problem: In a triangle that possesses a rectangular vertex O, there are given the lengths a and b of the two sides meeting in O.   Find the length c of the side opposite O.

We put the analogous problem: *In a tetrahedron that possesses a trirectangular vertex O, there are given the areas A, B, and C of the three faces meeting in O.   Find the area D of the face opposite O.*

We are required to express D in terms of A, B, and C.   It is natural to expect a formula analogous to Pythagoras' theorem

$$c^2 = a^2 + b^2$$

which solves the corresponding problem of plane geometry.   A high school boy guessed

$$D^3 = A^3 + B^3 + C^3$$

This is a clever guess; the change in the exponent corresponds neatly to the transition from 2 to 3 dimensions.

(3) *What is the unknown?* —The area of a triangle, D.

How can you find such an unknown? *How can you get this kind of thing?* —The area of a triangle can be computed if the three sides are known, by Heron's formula.   The area of our triangle is D.   Let a, b,

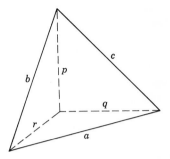

Fig. 2.3. Pythagoras in space.

and c denote the lengths of the sides, and set $s = (a + b + c)/2$; then

$$D^2 = s(s - a)(s - b)(s - c)$$

(This is a form of Heron's formula.) Let us label the sides of D in the figure; see Fig. 2.3.

Fine! But are the sides a, b, and c known? —No, but they are in right triangles; if the legs in these right triangles (labeled p, q, r in Fig. 2.3) were known, we could express a, b, and c:

$$a^2 = q^2 + r^2, \qquad b^2 = r^2 + p^2, \qquad c^2 = p^2 + q^2$$

That is good; but are p, q, and r themselves known? —No, but they are connected with the data, the areas A, B, and C:

$$\tfrac{1}{2}qr = A, \qquad \tfrac{1}{2}rp = B, \qquad \tfrac{1}{2}pq = C$$

That is right; but did you achieve anything useful? —I think I did. I now have 7 unknowns

$$D$$
$$a, \quad b, \quad c$$
$$p, \quad q, \quad r$$

but also a system of 7 equations to determine them.

(4) There is nothing wrong with our foregoing reasoning, under (3). We have attained the goal set by Descartes' Rule (freely rendered in sect. 2.3(3)): we have obtained a system with as many equations as there are unknowns. There is just one thing: the number 7 may seem too high, to solve 7 equations with 7 unknowns may appear as too much trouble. And Heron's formula may not look too inviting.

If we feel so, we may prefer a new start.

What is the unknown? —The area of a triangle, D.

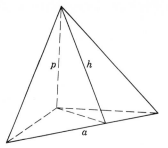

Fig. 2.4. A new departure.

How can you find such an unknown? *How can you get this kind of thing?* —The most familiar way to compute the area of a triangle is

$$D = \frac{ah}{2}$$

where a is the base, and h the altitude, of the triangle with area D; let us introduce h into the figure. See Fig. 2.4.

Yes, we have seen a before; but what about h? —The height h of the triangle with area D can be computed from a suitable triangle, I hope. In fact, intersect the tetrahedron with a plane passing through h and the trirectangular vertex. The intersection is a right triangle, its hypotenuse is h, one of its legs is p which we have seen before, and the other leg, say k, is the altitude perpendicular to the side a in the triangle with area A. Therefore,

$$h^2 = k^2 + p^2$$

Very good! But what about k? —We can get it somehow. In fact, express the area of the triangle in which, as I have just said, k is an altitude in two different manners:

$$\tfrac{1}{2}ak = A$$

Have you as many equations as you have unknowns? —I also have the former equations, and I have no time to count. I now see my way, I think. Let me just combine what is before me:

$$
\begin{aligned}
4D^2 &= a^2h^2 \\
&= a^2(k^2 + p^2) \\
&= 4A^2 + a^2p^2 \\
&= 4A^2 + (r^2 + q^2)p^2 \\
&= 4A^2 + (rp)^2 + (pq)^2 \\
&= 4A^2 + 4B^2 + 4C^2
\end{aligned}
$$

Let me bring together the beginning and the end and get rid of that super-fluous factor 4. Here it is:

$$D^2 = A^2 + B^2 + C^2$$

This result is, in fact, closely analogous to Pythagoras' theorem. That guess with the exponent 3 was clever—it turned out wrong, but this is not surprising. What is surprising is that the guess came so close to the truth.

It may be quite instructive to compare the two foregoing approaches to the same problem; they differ in various respects.

And could you imagine a different analogue to Pythagoras' theorem?

2.6. An example from physics

We start from the following question.

An iron sphere is floating in mercury. Water is poured over the mercury and covers the sphere. Will the sphere sink, rise, or remain at the same depth?

We have to compare two situations. In both cases the lower part of the iron sphere is immersed in (is under the level of) mercury. The upper part of the sphere is surrounded by air (or vacuum) in the first situation, and by water in the second situation. In which situation is the upper part (the one over the level of the mercury) a greater fraction of the whole volume?

This is a purely qualitative question. Yet we can give it a quantitative twist which renders it more precise (and accessible to algebra): *Compute the fraction of the volume of the sphere that is over the level of the mercury for both situations.*

(1) We can give a plausible answer to the qualitative question by purely intuitive reasoning, just by visualizing a *continuous transition* from one proposed situation to the other. Let us imagine that the fluid poured over the mercury and surrounding the upper part of the iron sphere *changes its density continuously.* To begin with, this imaginary fluid has density zero (we have just vacuum). Then the density increases; it soon attains the density of the air, and after a while the density of water. If you do not see yet how this change affects the floating sphere, let the *density increase still further.* When the density of that imaginary fluid attains the density of iron, the sphere must rise clear out of the mercury. In fact, if the density increased further ever so little, the sphere should pop up and emerge somewhat from that imaginary fluid.

It is natural to suppose that the position of the floating sphere, as the density of the imaginary fluid covering it, changes all the time in the *same direction.* Then we are driven to the conclusion that, in the transition from covering vacuum or air to covering water, the sphere will *rise.*

(2) In order to answer the quantitative question, we need the numerical values of the three specific gravities involved which are

	1.00	13.60	7.84
for	water	mercury	iron

respectively. Yet it is more instructive to substitute letters for these numerical data. Let

$$a \qquad b \qquad c$$

denote the specific gravity of the

upper fluid   lower fluid   floating solid

respectively. Let v denote the (given) total volume of the floating solid, x the fraction of v that is over the level separating the two fluids, and y the fraction under that level; see Fig. 2.5. Our data are a, b, c, and v, our unknowns x and y. It is understood that

$$a < c < b$$

We may express the total volume of the floating body in two different ways:

$$x + y = v$$

Now, we cannot proceed beyond this point unless we know the pertinent physical facts. The *relevant knowledge* that we should possess is the law of Archimedes which is usually expressed as follows: the floating body is buoyed up by a vertical force equal in magnitude to the weight of the displaced fluid. The sphere that we are considering displaces fluid in two layers. The weights of the displaced quantities are

	ax	and	by
in the	upper layer	and	lower layer

respectively. These two upward vertical forces must jointly balance the

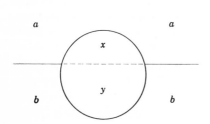

Fig. 2.5. Floating in two fluids.

weight of the floating sphere which we can, therefore, express in two different ways·

$$ax + by = cv$$

Now, we have obtained a system of two equations for our two unknowns x and y. Solving this system, we obtain

$$x = \frac{b - c}{b - a} v, \qquad y = \frac{c - a}{b - a} v$$

(3) Let us return to the original statement of the problem. In the first situation, if there is vacuum over the mercury

$$a = 0, \qquad b = 13.60, \qquad c = 7.84$$

which yield for the fraction of the iron sphere's volume over the level of mercury

$$x = 0.423v$$

In the second situation, when there is water over the mercury,

$$a = 1.00, \qquad b = 13.60, \qquad c = 7.84$$

which yield $\qquad\qquad x = 0.457v$

and the latter fraction is larger, which agrees with the conclusion of our intuitive reasoning.

The general formula (in letters) is, however, more interesting than any particular numerical result that we can derive from it. Especially, it fully substantiates our intuitive reasoning. In fact, keep b, c, and v constant and let a (the density of the upper layer) increase from

$$a = 0 \qquad \text{to} \qquad a = c$$

Then the denominator $b - a$ of x decreases steadily and so x, the fraction of v over the level of the mercury, increases steadily from

$$x = \frac{b - c}{b} v \qquad \text{to} \qquad x = v$$

2.7. An example from a puzzle

How can you make two squares from five? Fig. 2.6 shows a sheet of paper that has the shape of a cross; it is made up of five equal squares. Cut this sheet along a straight line in two pieces, then cut one of the pieces along another straight line again in two, so that the resulting three pieces, suitably fitted together, form two juxtaposed squares.

The cross in Fig. 2.6 is highly symmetric (it has a center of symmetry and four lines of symmetry). The two juxtaposed squares form a rectangle the length of which is twice the width. It is understood that the

Fig. 2.6. Two from five?

three pieces into which the cross will be divided should fill up this rectangle without overlapping.

Could you solve a part of the problem?   Obviously, the area of the desired rectangle is equal to the area of the given cross, and so it equals $5a^2$ if a denotes the side of one of the five squares forming the cross.   Yet, having obtained its area, we can also find the sides of the rectangle.   Let x denote the length of the rectangle; then its width is $x/2$.   Express the area of the rectangle in two different ways; we obtain

$$x \cdot \frac{x}{2} = 5a^2$$

or $$x^2 = 10a^2$$

from which we can find both sides of the rectangle.

We now have sufficient information about the rectangle, its shape and size, but the proposed puzzle is not yet solved: we still have to locate the two cuts in the cross.   Yet the expression for x obtained above may yield a hint, especially if we write it in this form:

$$x^2 = 9a^2 + a^2$$

With this indication, I leave the solution to the reader.

We can derive some useful suggestions from the foregoing treatment of the puzzle.

First, it shows that algebra can be useful even when it cannot solve the problem completely: it can solve a part of the problem and the solution of that part can facilitate the remaining work.

Second, the procedure that we have employed may impress us with its peculiar *expanding* pattern.   First, we have obtained only a small part of the solution: the area of the desired rectangle.   We have used, however, this small part to obtain a bigger part: the sides of the rectangle, and hence complete information about the rectangle.   Now, we are trying to use this bigger part to obtain a still bigger part which we may use afterwards, we hope, to obtain the full solution.

2.8. Puzzling examples

The problems that we have considered so far in this chapter are "reasonable." We are inclined to regard a problem as reasonable if its solution is uniquely determined. If we are seriously concerned with our problem, we wish to know (or guess) as early as possible whether it is reasonable or not. And so, from the outset, we may ask ourselves: *Is it possible to satisfy the condition? Is the condition sufficient to determine the unknown? Or is it insufficient? Or redundant? Or inconsistent?*

These questions are important.[4] We postpone a general discussion of their role, but it will be appropriate to look here at a couple of examples.

(1) *A man walked five hours, first along a level road, then up a hill, then he turned round and walked back to his starting point along the same route. He walks* 4 *miles per hour on the level,* 3 *uphill, and* 6 *downhill. Find the distance walked.*[5]

Is this a reasonable problem? Are the data *sufficient to determine the unknown? Or are they insufficient? Or redundant?*

The data seem to be *insufficient*: some information about the extent of the nonlevel part of the route seems to be lacking. If we knew how much time the man spent walking uphill, or downhill, there would be no difficulty. Yet without such information the problem appears indeterminate.

Still, let us try. Let

x stand for the total distance walked,
y for the length of the uphill walk.

The walk had four different phases:

$$\text{level,} \qquad \text{uphill,} \qquad \text{downhill,} \qquad \text{level.}$$

Now we can easily express the total time spent in walking in two different ways:

$$\frac{x/2 - y}{4} + \frac{y}{3} + \frac{y}{6} + \frac{x/2 - y}{4} = 5$$

Just one equation between two unknowns—it is insufficient. Yet, when we collect the terms, the coefficient of y turns out to be 0, and there remains

$$\frac{x}{4} = 5$$

$$x = 20$$

And so the data are sufficient to determine x, the only unknown required

---

[4] Cf. HSI, p. 122: Is it possible to satisfy the condition?
[5] Cf. "Knot I" of "A Tangled Tale" by Lewis Carroll.

by the statement of the problem. Hence, after all, the problem is not indeterminate: we were wrong.

(2) We were wrong, there is no denying, but we suspect that the author of the problem took pains to mislead us by a tricky choice of those numbers 3, 6, and 4. To get to the bottom of his trick, let us substitute for the numbers

$$3, \qquad\qquad 6, \qquad\qquad 4$$

the letters $\qquad\quad u, \qquad\qquad v, \qquad\qquad w$

which stand for the pace of the walk

$$\text{uphill}, \qquad \text{downhill}, \qquad \text{on the level},$$

respectively. We should reread the problem, with the letters just introduced substituted for the original numbers, and then express the total time spent in walking in two different ways, using the appropriate letters:

$$\frac{x/2 - y}{w} + \frac{y}{u} + \frac{y}{v} + \frac{x/2 - y}{w} = 5$$

or

$$\frac{x}{w} + \left(\frac{1}{u} + \frac{1}{v} - \frac{2}{w}\right) y = 5$$

We cannot determine x from this equation, unless the coefficient of y vanishes. And so the problem *is* indeterminate, unless

$$\frac{1}{w} = \frac{1}{2}\left(\frac{1}{u} + \frac{1}{v}\right)$$

If, however, the three rates of walking are chosen at random, they do not satisfy this relation, and so the problem *is* indeterminate. We were put in the wrong by a vicious trick!

(We can express the critical relation also by the formula

$$w = \frac{2uv}{u + v}$$

or by saying that the pace on the level is the harmonic mean of the uphill pace and the downhill pace.)

(3) *Two smaller circles are outside each other, but inside a third, larger circle. Each of these three circles is tangent to the two others and their centers are along the same straight line. Given r, the radius of the larger circle, and t, that piece of the tangent to the two smaller circles in their common point that lies within the larger circle. Find the area that is inside the larger circle but outside the two smaller circles. See Fig. 2.7.*

Is this a reasonable problem? Are the data *sufficient to determine the unknown? Or are they insufficient? Or redundant?*

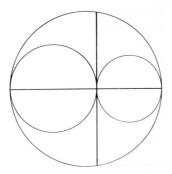

Fig. 2.7. Two data.

The problem seems perfectly reasonable. To determine the configuration of the three circles, it is both necessary and sufficient to know the radii of the two smaller circles, and any two independent data will be just as good. Now, the given r and t are obviously independent: we can vary one without changing the other (except for the inequality $t \leqq 2r$ which we may take for granted). Yes, the two data r and t seem to be just sufficient, neither insufficient nor redundant.

Therefore, let us settle down to work. Let A stand for the required area, x and y for the radii of the two smaller circles. Obviously

$$A = \pi r^2 - \pi x^2 - \pi y^2$$
$$2r = 2x + 2y$$

We have here two equations for our three unknowns, A, x, and y. In order to obtain a third equation, consider the right triangle inscribed in the larger circle, the base of which passes through the three centers and the opposite vertex of which is one of the endpoints of the segment t. The altitude in this triangle, drawn from the vertex of the right angle, is $t/2$; this altitude is a mean proportional (Euclid VI 13):

$$\left(\frac{t}{2}\right)^2 = 2x \cdot 2y$$

Now we have three equations. We rewrite the last two:

$$(x + y)^2 = r^2$$

$$2xy = \frac{t^2}{8}$$

Subtraction yields $x^2 + y^2$, and substitution into the first equation yields

$$A = \frac{\pi t^2}{8}$$

The data turned out to be *redundant*: of the two data, t and r, only the first is really needed, not the second. We were wrong again.

The curious relation underlying the example just discussed was observed by Archimedes; see his *Works* edited by T. L. Heath, pp. 304–305.

Examples and Comments on Chapter 2

First Part

2.1. Bob has three dollars and one half in nickels and dimes, fifty coins altogether. How many nickels has Bob, and how many dimes? (Have you seen the same problem in a slightly different form?)

2.2. Generalize the problem of sect. 2.4(1) by passing from "numbers" to "letters" and considering several filling and emptying pipes.

2.3. Devise some other interpretation for the equation set up in sect. 2.4(2).

2.4. Find further checks for the solution of the flight problem of sect. 2.4(3).

2.5. In the "mixture problem" of sect. 2.4(4) substitute the letters

$$a \quad b \quad c \quad v$$

for the numerical data

$$90 \quad 60 \quad 72 \quad 50$$

respectively. Read the problem after this substitution and set up the equations. Do you recognize them?

2.6. Fig. 2.8 (which is different from, but related to, Fig. 2.1) shows another configuration frequently seen in Gothic tracery.

Find the center of the circle that touches four circular arcs forming a "curvilinear quadrilateral."

Two arcs have the radius AB, the center of one is A, that of the other B.

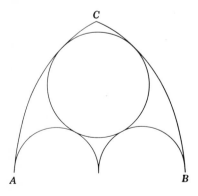

Fig. 2.8. From a Gothic window.

Two semicircles have the radius $AB/4$, the center of each lies on the line AB, one starts from the point A, the other from the point B, both end in the midpoint of the line AB where they are tangent to each other.

2.7. Carry through the plan devised in sect. 2.5(3); it should lead you to the same simple expression for D^2 in terms of A, B, and C that we have obtained in sect. 2.5(4) by other means.

2.8. Compare the approaches of sect. 2.5(3) and 2.5(4). (Emphasize general viewpoints.)

2.9. Find the volume V of a tetrahedron that has a trirectangular vertex O, being given the areas A, B, and C of the three faces meeting in O.

2.10. *An analogue to Heron's theorem.*   Find the volume V of a tetrahedron that has a trirectangular vertex, being given the lengths a, b, and c of the sides of the face opposite the trirectangular vertex.

(If we introduce the quantity

$$S^2 = \frac{a^2 + b^2 + c^2}{2}$$

into the expression of V in an appropriate, symmetric way, the result assumes a form somewhat similar to Heron's formula.)

2.11. *Another analogue to Pythagoras' theorem.*   Find the length d of the diagonal of a box (a rectangular parallelepiped) being given p, q, and r, the length, the width, and the height of the box.

2.12. *Still another analogue to Pythagoras' theorem.*   Find the length d of the diagonal of a box, being given a, b, and c, the lengths of the diagonals of three faces having a corner in common.

2.13. *Another analogue to Heron's theorem.*   Let V denote the volume of a tetrahedron, a, b, and c the lengths of the three sides of one of its faces, and assume that each edge of the tetrahedron is equal in length to the opposite edge. Express V in terms of a, b, and c.

2.14. Check the result of ex. 2.10 and that of ex. 2.13 by examining the degenerate case in which V vanishes.

2.15. Solve the puzzle proposed in sect. 2.7. (The sides x and $x/2$ should result from the cuts—but how can you fit a segment of length x into the cross?)

2.16. Fig. 2.9 shows a sheet of paper of peculiar shape: it is a rectangle with a rectangular hole. The sides of the outer rectangular boundary measure 9 and 12, those of the inner 1 and 8 units, respectively. Both rectangular boundaries have the same center and their corresponding sides are parallel. Cut this sheet along just two lines in two pieces that, fitted together, form a complete square.

(*a*) *Could you solve a part of the problem?*   How long is the side of the desired square?

(*b*) *Take the problem as solved.*   Imagine that the sheet is already cut into

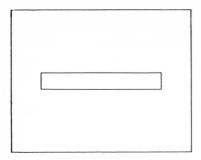

Fig. 2.9. By two cuts a square.

two pieces, the "left piece" and the "right piece." You keep the left piece where it is, and move the right piece into the desired position (where, with the other, it forms a complete square). Knowing the answer to (*a*), what kind of motion do you expect?

(*c*) *Guess a part of the solution.* The given sheet is symmetric with respect to its center and also with respect to two axes perpendicular to each other. Which kind of symmetry do you expect it to retain when cut along the two required lines?

Second Part

Some of the following examples are grouped according to subject matter which is hinted by an indication in front of the first example of the group (*Miscellaneous, Plane Geometry, Solid Geometry*, etc.) Some examples are followed by the name of Newton or Euler in parentheses; these are taken from the following sources, respectively:

Universal Arithmetick: or, a Treatise of Arithmetical Composition and Resolution. Written in Latin by Sir Isaac Newton. Translated by the late Mr. Ralphson. London, 1769. (Examples marked "After Newton" are derived from the same source, but some change is introduced into the formulation or into the numerical data.)

Elements of Algebra. By Leonard Euler. Translated from the French. London, 1797. (In fact, Euler's Algebra was originally written in German.)

Isaac Newton (1643–1727) is regarded by many as the greatest man of science who ever lived. His work encompasses the principles of mechanics, the theory of universal gravitation, the differential and integral calculus, theoretical and experimental optics, and several minor items each of which would be sufficient to secure him a place in the history of science. Leonard Euler (1707–1783) is also one of the very great; he left his traces on almost every branch of mathematics and on several branches of physics; he contributed more than anybody else to the development of the calculus discovered by Newton and Leibnitz. Observe that such great men did not find it beneath their dignity to explain and illustrate at length the application of equations to the solution of "word problems."

2.17. *Miscellaneous.* A mule and an ass were carrying burdens amounting to some hundred weight. The ass complained of his, and said to the mule: "I need only one hundred weight of your load, to make mine twice as heavy as yours." The mule answered: "Yes, but if you gave me a hundred weight of yours, I should be loaded three times as much as you would be." How many hundred weight did each carry? (Euler)

2.18. When Mr. and Mrs. Smith took the airplane, they had together 94 pounds of baggage. He paid $1.50 and she paid $2 for excess weight. If Mr. Smith made the trip by himself with the combined baggage of both of them, he would have to pay $13.50. How many pounds of baggage can one person take along without charge?

2.19. A father who has three sons leaves them 1600 crowns. The will precises, that the eldest shall have 200 crowns more than the second, and the second shall have 100 crowns more than the youngest. Required the share of each. (Euler)

2.20. A father leaves four sons, who share his property in the following manner:

The first takes the half of the fortune, minus 3000 livres.

The second takes the third, minus 1000 livres.

The third takes exactly the fourth of the property.

The fourth takes 600 livres and the fifth part of the property.

What was the whole fortune, and how much did each son receive? (Euler)

2.21. A father leaves at his death several children, who share his property in the following manner:

The first receives a hundred crowns and the tenth part of what remains.

The second receives two hundred crowns and the tenth part of what remains.

The third takes three hundred crowns and the tenth part of what remains.

The fourth takes four hundred crowns and the tenth part of what remains, and so on.

Now it is found at the end that the property has been divided equally among all the children. Required, how much it was, how many children there were, and how much each received. (Euler)

2.22. Three persons play together; in the first game, the first loses to each of the other two as much money as each of them has. In the next, the second person loses to each of the other two as much money as they have already. Lastly, in the third game, the first and second person gain each from the third as much money as they had before. They then leave off and find that they have all an equal sum, namely, 24 louis each. Required, with how much money each sat down to play. (Euler)

2.23. Three Workmen can do a Piece of Work in certain Times, viz. *A* once in 3 Weeks, *B* thrice in 8 Weeks, and *C* five Times in 12 Weeks. It is desired to know in what Time they can finish it jointly. (Newton)

2.24. The Forces of several Agents being given, to determine the Time wherein they will jointly perform a given Effect. (Newton)

2.25. One bought 40 Bushels of Wheat, 24 Bushels of Barley, and 20 Bushels of Oats together for 15 Pounds 12 Shillings.

Again, he bought of the same Grain 26 Bushels of Wheat, 30 Bushels of Barley, and 50 Bushels of Oats together for 16 Pounds.

And thirdly, he bought the like Kind of Grain, 24 Bushels of Wheat, 120 Bushels of Barley, and 100 Bushels of Oats together for 34 Pounds.

It is demanded at what Rate a Bushel of each of the Grains ought to be valued. (Newton)

2.26. (Continued) Generalize.

2.27. If 12 Oxen eat up $3\frac{1}{3}$ Acres of Pasture in 4 Weeks, and 21 Oxen eat up 10 Acres of like Pasture in 9 Weeks; to find how many Oxen will eat up 24 Acres in 18 Weeks. (Newton)

2.28. *An Egyptian problem.* We take a problem from the Rhind Papyrus which is the principal source of our knowledge of ancient Egyptian mathematics. In the original text, the problem is about hundred loaves of bread which should be divided between five people, but a major part of the condition is not expressed (or not clearly expressed); the solution is attained by "groping": by a guess, and a correction of the first guess.[6]

Here follows the Egyptian problem reduced to abstract form and modern terminology; the reader should go one step further and reduce it to equations: An arithmetic progression has five terms. The sum of all five terms equals 100, the sum of the three largest terms is seven times the sum of the two smallest terms. Find the progression.

2.29. A geometric progression has three terms. The sum of these terms is 19 and the sum of their squares is 133. Find the terms. (After Newton.)

2.30. A geometric progression has four terms. The sum of the two extreme terms is 13, the sum of the two middle terms is 4. Find the terms. (After Newton.)

2.31. Some merchants have a common stock of 8240 crowns; each contributes to it forty times as many crowns as there are partners; they gain with the whole sum as much per cent as there are partners; dividing the profit, it is found that, after each has received ten times as many crowns as there are partners, there remain 224 crowns. Required the number of partners. (Euler)

2.32. *Plane geometry.* Inside a square with side a there are five nonoverlapping circles with the same radius r. One circle is concentric with the square and touches the four other circles each of which touches two sides of the square (is pushed into a corner). Express r in terms of a.

2.33. *Newton on setting up equations in geometric problems.* If the Question be of an Isosceles Triangle inscribed in a Circle, whose Sides and Base are to be compared with the Diameter of the Circle, this may either be proposed of

[6] Cf. J. R. Newman, *The World of Mathematics*, vol. 1, pp. 173–174.

the Investigation of the Diameter from the given Sides and Base, or of the Investigation of the Basis from the given Sides and Diameter; or lastly, of the Investigation of the Sides from the given Base and Diameter; but however it be proposed, it will be reduced to an Equation by the same...Analysis. (Newton)

Let d, s, and b stand for the length of the diameter, that of the side, and that of the base, respectively (so that the three sides of the triangle are of length s, s, and b, respectively), and find an equation connecting d, s, and b which solves all three problems: the one in which d, the other in which b, and the third in which s is the unknown. (There are always two data.)

2.34. (Continued) Examine critically the equation obtained as answer to ex. 2.33. (a) Are all three problems equally easy? (b) The equation obtained yields positive values in the three cases mentioned (for d, b, and s, respectively) only under certain conditions: do these conditions correctly correspond to the geometric situation?

2.35. The four points G, H, V, and U are (in this order) the four corners of a quadrilateral. A surveyor wants to find the length $UV = x$. He knows the length $GH = l$ and measures the four angles

$$\angle GUH = \alpha, \qquad \angle HUV = \beta, \qquad \angle UVG = \gamma, \qquad \angle GVH = \delta.$$

Express x in terms of α, β, γ, δ, and l.

(Remember ex. 2.33 and follow Newton's advice: choose those "Data and Quaesita by which you think it is most easy for you to make out your Equation.")

2.36. The Area and the Perimeter of a right-angled Triangle being given, to find the Hypothenuse. (Newton)

2.37. Having given the Altitude, Base and Sum of the Sides, to find the Triangle. (Newton)

2.38. Having given the Sides of any Parallelogram and one of the Diagonals, to find the other Diagonal. (Newton)

2.39. The triangle with the sides a, a, and b is isosceles. Cut off from it two triangles, symmetric to each other with respect to the altitude perpendicular to the base b, so that the remaining symmetric pentagon is *equilateral*. Express the side x of the pentagon in terms of a and b.

(This problem was discussed by Leonardo of Pisa, called Fibonacci, with the numerical values $a = 10$ and $b = 12$.)

2.40. A hexagon is equilateral, its sides are all of the same length a. Three of its angles are right angles; they alternate with three obtuse angles. (If the hexagon is $ABCDEF$, the angles at the vertices A, C, and E are right angles, those at the vertices B, D, and F obtuse.) Find the area of the hexagon.

2.41. An equilateral triangle is inscribed in a larger equilateral triangle so that corresponding sides of the two triangles are perpendicular to each other.

Thus the whole area of the larger triangle is divided into four pieces each of which is a fraction of the whole area. Which fraction?

2.42. Divide a given triangle by three straight cuts into seven pieces four of which are triangles (and the remaining three pentagons). One of the triangular pieces is included by the three cuts, each of the three other triangular pieces is included by a certain side of the given triangle and two cuts. Choose the three cuts so that the four triangular pieces turn out to be congruent. Which fraction of the area of the given triangle is the area of a triangular piece in this dissection?

(It may be advantageous to examine first a particular shape of the given triangle for which the solution is particularly easy.)

2.43. The point P is so located in the interior of a rectangle that the distance of P from a corner of the rectangle is 5 yards, from the opposite corner 14 yards, and from a third corner 10 yards. What is the distance of P from the fourth corner?

2.44. Given the distances a, b, and c of a point in the plane of a square from three vertices of the square; a and c are distances from opposite vertices.

(I) Find the side s of the square.

(II) Test your result in the following four particular cases:

(1) $a = b = c$
(2) $b^2 = 2a^2 = 2c^2$
(3) $a = 0$
(4) $b = 0$.

2.45. Pennies (equal circles) are arranged in a regular pattern all over a very, very large table (the infinite plane). We examine two patterns.

In the first pattern, each penny touches four other pennies and the straight lines joining the centers of the pennies in contact dissect the plane into equal squares.

In the second pattern, each penny touches six other pennies and the straight lines joining the centers of the pennies in contact dissect the plane into equal equilateral triangles.

Compute the percentage of the plane covered by pennies (circles) for each pattern.

(For the first pattern see Fig. 3.9, for the second Fig. 3.8.)

2.46. *Solid geometry.* Inside a cube (a cubical box) with edge a there are nine nonoverlapping spheres (nine balls packed into the box) with the same radius r. One sphere is concentric with the cube and touches the eight other spheres (the balls are tightly packed), each of which touches three faces of the cube (is pushed into a corner). Express r in terms of a.

(Or a in terms of r: make a box when you have the balls. There is an analogous problem in plane geometry, see ex. 2.32: can you use its result or method?)

2.47. Devise a problem of solid geometry analogous to ex. 2.43.

2.48. A pyramid is called "regular" if its base is a regular polygon and the foot of its altitude is the center of its base.

All five faces of a regular pyramid with square base are of the same area. Given the height h of the pyramid, find the area of its surface.

2.49. (Continued) There is some analogy between a regular pyramid and an isosceles triangle; at any rate, if the number of the lateral faces of the pyramid is given, both figures, the solid and the plane, depend on two data.

Devise further problems about regular pyramids.

2.50. Devise a proposition of solid geometry analogous to the result of ex. 2.38. (Ex. 2.12 may serve as a stepping stone to a generalization.)

2.51. Find the area of the surface of the tetrahedron considered in ex. 2.13, being given a, b, and c. (Do you see some analogy?)

2.52. Of twelve congruent equilateral triangles eight are the faces of a regular octahedron and four the faces of a regular tetrahedron. Find the ratio of the volume of the octahedron to the volume of the tetrahedron.

2.53. A triangle rotating first about its side a, then about its side b, and finally about its side c, generates three solids of revolution. Find the ratio of the volumes and that of the surface-areas of these three solids.

2.54. *An inequality.* A rectangle and an isosceles trapezoid are in the relative situation shown in Fig. 2.10: they have a common (vertical) line of symmetry, the same height h and the same area; if $2a$ and $2b$ are the lengths of the lower and upper base of the trapezoid, respectively, the base of the rectangle is of length $a + b$. Rotated about the common line of symmetry, the rectangle describes a cylinder and the trapezoid the frustum of a cone. Which one of these two solids has the greater volume? (Your answer may be suggested by geometry, but should be proved by algebra.)

2.55. *Spherometer.* There are four points A, B, C, and D on the surface of

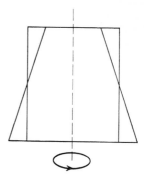

Fig. 2.10. Rotate it!

a sphere. The points A, B, and C form an equilateral triangle with side a.
A perpendicular drawn from D to the plane of $\triangle ABC$ has the length h, and
its foot is the center of $\triangle ABC$.

Given a and h, compute the radius R of the sphere.

(This geometric situation underlies the construction of the spherometer
which is an instrument to determine the curvature of a lens. On the spherometer, A, B, and C are the endpoints of three fixed parallel "legs," whereas the
endpoint of a fourth, movable "leg" is screwed into the position D and the
distance h is measured by the revolutions of the screw.)

2.56. *Graphic time table.* In problems about several objects (material
points) moving along the same path it is often advantageous to introduce a
rectangular coordinate system in which the abscissa represents t, the time, and
the ordinate represents s, the distance, measured along the path from a fixed
point. To show the use of this device, we reconsider the problem that we
have treated in detail in sect. 2.4(3).

We measure the time t and the distance s from the starting time and the
starting point of the airplane, respectively. Thus, when the airplane has
traveled t hours on its outgoing flight, its distance from its starting point is

$$s = (v - w)t$$

This equation, with fixed v and w and variable s and t, is represented in our
coordinate system by a straight line with slope $v - w$ (the velocity) that passes
through the origin [the point $(0, 0)$ which represents the start of the airplane].
On the returning flight, the distance s and the time t are connected by the
equation

$$s = -(v + w)(t - T)$$

of a straight line with slope $-(v + w)$ passing through the point $(T, 0)$ (which
represents the coming back of the airplane to its starting point at the prescribed time T).

The intersection of the two lines represents the point (in space and time)
that belongs both to the outgoing, and to the returning flight, where the airplane reverses its direction. At this point, both expressions are valid for s
simultaneously, and so

$$(v - w)t = -(v + w)(t - T)$$

This yields

$$t = \frac{(v + w)T}{2v}$$

and therefore (from either expression for s)

$$s = \frac{(v^2 - w^2)T}{2v}$$

as expression for the distance of the farthest point reached by the airplane.
This is the result we found in sect. 2.4(3) (with x instead of s).

In Fig. 2.11 (disregard the dotted segments) the flight of the airplane is

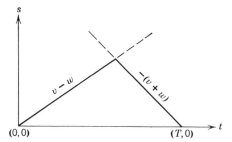

Fig. 2.11. Graphic timetable.

represented by a line consisting of two straight pieces; these pieces form an angle at the point whose ordinate represents the greatest distance reached by the airplane. The whole line tells the whole story of the flight; it shows where the airplane was at any given time and when it arrived at any given point; this line is appropriately called the *graphic time table* of the flight (of the motion considered).

2.57. Two Post-Boys A and B, at 59 Miles Distance from one another, set out in the Morning in order to meet. And A rides 7 Miles in two Hours, and B 8 Miles in three Hours, and B begins his Journey one Hour later than A; to find what Number of Miles A will ride before he meets B. (Newton)

2.58. (Continued) Generalize.

2.59. Al and Bill live at opposite ends of the same street. Al had to deliver a parcel at Bill's home, Bill one at Al's home. They started at the same moment; each walked at constant speed and returned home immediately after leaving the parcel at its destination. They met the first time face to face, coming from opposite directions, at the distance of a yards from Al's home and the second time at the distance of b yards from Bill's home.
(1) How long is the street?
(2) If $a = 300$ and $b = 400$ who walks faster?

2.60. Bob, Peter, and Paul travel together. Peter and Paul are good hikers; each walks p miles per hour. Bob has a bad foot and drives a small car in which two people can ride, but not three; the car covers c miles per hour. The three friends adopted the following scheme: they start together, Paul rides in the car with Bob, Peter walks. After a while, Bob drops Paul who walks on; Bob returns to pick up Peter, and then Bob and Peter ride in the car till they overtake Paul. At this point, they change: Paul rides and Peter walks just as they started and the whole procedure is repeated as often as necessary.
(1) How much progress (how many miles) does the company make per hour?
(2) Through which fraction of the travel time does the car carry just one man?

2.61. (Continued) Generalize: Bob, with his bad foot and small car, makes a similar arrangement with n friends, A, B, C,..., and L (instead of two) all walking at the pace p miles per hour.

(Draw a graphic time table for $n = 3$. Check the extreme cases $p = 0$, $p = c$, $n = 1$, $n = \infty$.)

2.62. A Stone falling down into a Well, from the Sound of the Stone striking the Bottom, to determine the Depth of the Well. (Newton)

You have to measure the time T between two moments: the first when you let the stone go, and the second when you hear it striking the bottom. You have to know too:

c the velocity of sound and

g the gravitational acceleration.

Being given T, c, and g, find d, the depth of the well.

2.63. To determine the Position of a Comet's Course that moves uniformly in a right Line from three Observations. (Newton)

Let O be the eye of the observer, A, B, and C the place of the comet in the first, the second, and the third observation, respectively. From observations, we know the angles

$$\angle AOB = \omega, \qquad \angle AOC = \omega'$$

and the times, t and t', between the first and the second, and between the first and the third, observations, respectively. From the assumption of uniform motion

$$\frac{AB}{AC} = \frac{t}{t'}$$

Being given ω, ω', t, and t', find $\beta = \angle ABO$.

(Express some trigonometric function of β, for instance, $\cot \beta$, in terms of ω, ω', t, and t'.)

2.64. *As many equations as unknowns.* Find x, y, and z from the system of three equations

$$3x - y - 2z = a$$
$$-2x + 3y - z = b$$
$$-x - 2y + 3z = c$$

a, b, and c are given.

(*Is it possible to satisfy the condition? Is the condition sufficient to determine the unknowns?*)

2.65. *More equations than unknowns.* Find three numbers p, q, and r so that the equation

$$x^4 + 4x^3 - 2x^2 - 12x + 9 = (px^2 + qx + r)^2$$

holds identically for variable x.

(This problem requires the "exact" extraction of a square root of a given polynomial of degree 4, which may be possible in the present case, yet usually it is not. Why not?)

2.66. Show that it is impossible to find (real or complex) numbers a, b, c, A, B, and C such that the equation

$$x^2 + y^2 + z^2 = (ax + by + cz)(Ax + By + Cz)$$

holds identically for independently variable x, y, and z.

2.67. *Fewer equations than unknowns.* A certain person buys hogs, goats, and sheep, to the number of 100, for 100 crowns; the hogs cost him $3\frac{1}{2}$ crowns a piece, the goats $1\frac{1}{3}$ crown, and the sheep $\frac{1}{2}$ a crown; how many had he of each? (Euler)

Euler solves this problem by a procedure which he calls *Regula Caeci* ("Blind Man's Rule") as follows. Let x, y, and z stand for the number of hogs, goats, and sheep bought, respectively; x, y, and z should be, of course, *positive integers*. Expressing first the total number, then the total price, of the animals bought, we obtain

$$x + y + z = 100$$
$$21x + 8y + 3z = 600$$

the second equation has been slightly (and advantageously) transformed. If we eliminate z and solve the resulting equation for y, we obtain

$$y = 60 - \frac{18x}{5}$$

whence we conclude that

$$\frac{x}{5} = t$$

must be a positive integer.

Finish the solution.

2.68. A coiner (we hope that the coins he makes are not counterfeit) has three kinds of silver, the first of 7 ounces, the second of $5\frac{1}{2}$ ounces, the third of $4\frac{1}{2}$ ounces fine per marc (a marc is 8 ounces), and he wishes to form a mixture of the weight of 30 marcs at 6 ounces (fine per marc): how many marcs of each sort must he take? (Euler; additions in parenthesis.)

It is understood that a solution *in integers* is required. A problem is called a *Diophantine problem*, when its condition admits only integral values for the unknowns.

2.69. There is a number (an integer) which yields a square if you add it to 100 and another square if you add it to 168. Which number is it?

2.70. Bob's stamp collection consists of three books. Two tenths of his stamps are in the first book, several sevenths in the second book, and there are 303 stamps in the third book. How many stamps has Bob? (Is the condition sufficient to determine the unknown?)

2.71. A certain make of ball point pen was priced 50 cents in the store opposite the high school but found few buyers. When, however, the store had reduced the price, the whole remaining stock was sold for $31.93. What was the reduced price? (Is the condition sufficient to determine the unknown?)

2.72. *Descartes' Rules.* The work of the great philosopher and mathematician René Descartes, quoted in sect. 2.1, is of particular importance for our study.

The "Rules" were found unfinished in Descartes' manuscripts and published posthumously. Descartes planned 36 sections, but the work consists actually of 18 sections in more or less finished form and of the summaries of three more; the rest was very probably never written. The first twelve sections treat of mental operations useful in solving problems, the next twelve discuss "perfectly understood" problems, and the last twelve sections were intended to deal with "imperfectly understood" problems.[7]

Each section begins with a "Rule," a terse advice to the reader, and the section motivates, explains, exemplifies, or states in greater detail, the idea summarized by the Rule. We shall quote any passage of the section following a certain Rule by giving the number of that Rule.[8]

The words of Descartes will be a valuable guide for us, but the reader would offend the memory of the originator of "Cartesian doubt" if he believed anything that Descartes said merely because Descartes said it. In fact, the reader should not believe what the present author says, or any other author says, or trust too much his own hasty impressions. After having given a fair hearing to the author, the reader should accept only such statements which he can see pretty clearly by himself or of which his well-digested experience can convince him. Doing so, he will act according to the spirit of Descartes' "Rules."

2.73. Strip the problem. We quote Descartes: *Strip the question of superfluous notions and reduce it to its simplest form.*[9] This advice is applicable to problems of all kinds and on all levels. Yet let us be specific. Take a usual type of classroom problem, a "rate problem" about motion [such as that discussed in sect. 2.4(3)]. The moving object may be described by the problem as a person, a car, a train, or an airplane. In fact, it makes little difference: in solving the problem on this primitive level, we actually treat the object as a material point moving uniformly along a straight path. Such simplification may be quite reasonable in some cases, ridiculous in other cases. It is certain, however, that we cannot avoid some degree of *simplification* or *abstraction* when we are reducing a problem about real objects to a mathematical problem. For, obviously, a mathematical problem deals with abstractions; it is concerned with real objects only indirectly by virtue of a previous passage from the concrete to the abstract.

Engineers and physicists, in handling their problems, have to devote serious thought to the question, how far should the abstraction and simplification go, which details can be neglected, which small effects disregarded. They have to avoid two opposite dangers: they should not render the mathematical task too formidable, yet they should not oversimplify the physical situation.

---

[7] Rule XII, pp. 429–430. "Perfectly understood problems" can be, "imperfectly understood problems" cannot be, immediately reduced to purely mathematical problems—this seems to be the most relevant difference.

[8] As it was done in footnote 7. We quote the standard edition of the *Œuvres de Descartes* by Charles Adam and Paul Tannery, vol. X, which contains the Latin original of the "Rules," "Regulae ad directionem ingenii" on pp. 359–469. Other quotations from Descartes refer to the same edition, but give also the volume.

[9] Rule XIII, p. 430.

We get a foretaste of their dilemma in dealing with the most primitive word problems. In fact, getting accustomed to a degree of simplification is one of the difficulties that the beginner has to master, and if this difficulty is never brought out into the open, it may become much worse.

There is a related difficulty. The problems proposed in textbooks tacitly assume certain simplifications; real rates may be variable, but all the rates considered in elementary textbooks are constant. The beginner has to become familiar with such tacit assumptions, he has to learn the proper *interpretation* of certain conventionally abbreviated formulations; also this point should be openly discussed, at least now and then.

(There is another related point which is so important that we must mention it, but so far away from our main line of inquiry that we must mention it briefly: We should relate to the extent of simplification and neglect that enters into the formulation of the problem the precision to which we carry the numerical computation of the unknown. We may sin by transgression or omission if we compute more or less decimals than the data warrant. There are few occasions to illustrate this important point on the elementary level, but those few occasions should not be missed.)

For an instructive and not too obvious illustration of some of the points here discussed see the author's paper no. 18 quoted in the Bibliography.

2.74. Relevant knowledge. Mobilization and organization. Obviously, we cannot translate a physical problem into equations unless we know (or rather assume as known) the pertinent physical facts. For instance, we could not have solved the problem treated in sect. 2.6 without the knowledge of the law of Archimedes.

In translating a geometrical problem into equations, we use pertinent geometrical facts. For instance, we may apply the theorem of Pythagoras, or the proportionality of sides in similar triangles, or expressions for areas or volumes, and so on.

If we do not possess the relevant knowledge, we cannot translate the problem into equations. Yet even if we have once acquired such knowledge, it may not be present to our mind in the moment we need it; or if, by some chance, it is present we may fail to recognize its utility for the purpose at hand. We can see here clearly a point which should be obvious: It is not enough to possess the needed relevant knowledge in some dormant state; we have to recall it when needed, revive it, *mobilize* it, and make it available for our purpose, adapt it to our problem, *organize* it.

As our work progresses, our conception of the problem continually changes: more and more auxiliary lines appear in the figure, more and more auxiliary unknowns appear in our equations, more and more materials are mobilized and introduced into the structure of the problem till eventually the figure is saturated, we have just as many equations as unknowns, and originally present and successively mobilized materials are merged into an organic whole. (Sect. 2.5(3) yields a good illustration.)

2.75. Independence and consistence. Descartes advises us to set up as many equations as there are unknowns.[10]   Let n stand for the number of unknowns, and $x_1, x_2, \ldots, x_n$ for the unknowns themselves; then we can write the desired system of equations in the form

$$r_1(x_1, x_2, \ldots, x_n) = 0$$
$$r_2(x_1, x_2, \ldots, x_n) = 0$$
$$\cdot \quad \cdot \quad \cdot \quad \cdot \quad \cdot \quad \cdot$$
$$r_n(x_1, x_2, \ldots, x_n) = 0$$

where $r_1(x_1, x_2, \ldots, x_n)$ indicates a polynomial in $x_1, x_2, \ldots, x_n$, and the left-hand sides of the following equations must be similarly interpreted.   Descartes advises us further to reduce this system of equations to one final equation.[11] This is possible "in general" (usually, in the regular case,...) and "in general" the system has a solution (a system of numerical values, real or complex, for $x_1, x_2, \ldots, x_n$ satisfying simultaneously the n equations) and only a finite number of solutions (this number depends on the degree of the final equation).

Yet there are exceptional (irregular) cases; we cannot discuss them here in full generality, but let us look at a simple example.

We consider a system of three linear equations with three unknowns:

$$a_1 x + b_1 y + c_1 z + d_1 = 0$$
$$a_2 x + b_2 y + c_2 z + d_2 = 0$$
$$a_3 x + b_3 y + c_3 z + d_3 = 0$$

x, y, z are the unknowns, the twelve symbols $a_1, b_1, \ldots, d_3$ represent given real numbers.   We assume that a_1, b_1, and c_1 are not all equal to 0, and similarly for a_2, b_2, c_2 and a_3, b_3, c_3.   Under these assumptions, each of the three equations is represented by a plane if we consider x, y, and z as rectangular coordinates in space; and so the system of three equations is represented by a configuration of three planes.

Concerning the system of our three linear equations, we distinguish various cases.

(1) There exists no solution, that is, no system of three real numbers x, y, z satisfying all three equations simultaneously.   In this case we say that the equations are *incompatible*, and their system is *inconsistent* or *self-contradictory*.

(2) There are infinitely many solutions; then we say that the system is *indeterminate*.   This will be the case if all triplets of numbers x, y, and z that satisfy two of our equations, satisfy also the third, in which case we say that this third equation is *not independent* of the other two.

(3) There is just one solution: the equations are *independent*, the system is *consistent* and *determinate*.

Visualize the corresponding cases for the configuration of the three planes and describe them.

2.76. *Unique solution.   Anticipation.*   If a chess problem or a riddle has more than one solution, we regard it as imperfect.   In general, we seem to

[10] Rule XIX, p. 468.
[11] Rule XXI, p. 469.

have a natural preference for problems with a unique solution, they may appear to us as the only "reasonable" or the only "perfect" problems. Also Descartes seems to have shared this preference; he says: "*A perfect question, as we wish to define it, is completely determinate and what it asks can be derived from the data.*"[12]

Is the solution of our problem unique? *Is the condition sufficient to determine the unknown?* We often ask these questions fairly soon (and it is advisable to ask them so) when we are working on a problem. In asking them so early, we do not really need, or expect, a final answer (which will be given by the complete solution), only a preliminary answer, an *anticipation*, a guess (which may deepen our understanding of the problem). This preliminary answer often turns out to be right, but now and then we may fall into a trap as the examples of sect. 2.8 show.[13]

By the way, the unknown may be obtained as a root of an equation of degree n with n different roots where $n > 1$, and the solution may still be unique, if the condition requires a real, or a positive, or an integral value for the unknown and the equation in question has just one root of the required kind.

2.77. *Why word problems?* I hope that I shall shock a few people in asserting that the most important single task of mathematical instruction in the secondary schools is to teach the setting up of equations to solve word problems. Yet there is a strong argument in favor of this opinion.

In solving a word problem by setting up equations, the student *translates* a real situation into mathematical terms: he has an opportunity to experience that mathematical concepts may be related to realities, but such relations must be carefully worked out. Here is the first opportunity afforded by the curriculum for this basic experience. This first opportunity may be also the last for a student who will not use mathematics in his profession. Yet engineers and scientists who will use mathematics professionally, will use it mainly to translate real situations into mathematical concepts. In fact, an engineer makes more money than a mathematician and so he can hire a mathematician to solve his mathematical problems for him; therefore, the future engineer need not study mathematics to *solve* problems. Yet, there is one task for which the engineer cannot fully rely on the mathematician: the engineer must know enough mathematics to *set up* his problems in mathematical form. And so the future engineer, when he learns in the secondary school to set up equations to solve "word problems," has a first taste of, and has an opportunity to acquire the attitude essential to, his principal professional use of mathematics.

2.78. *More problems.* Devise some problems similar to, but different from, the problems proposed in this chapter—especially such problems as you can solve.

---

[12] Rule XIII, p. 431.
[13] For more examples of this kind see MPR, vol. 1, pp. 190–192, and pp. 200–202, problems 1–12.

CHAPTER 3

RECURSION

3.1. The story of a little discovery

There is a traditional story about the little Gauss who later became the great mathematician Carl Friedrich Gauss.   I particularly like the following version which I heard as a boy myself, and I do not care whether it is authentic or not.

"This happened when little Gauss still attended primary school.   One day the teacher gave a stiff task: To add up the numbers 1, 2, 3, and so on, up to 20.   The teacher expected to have some time for himself while the boys were busy doing that long sum.   Therefore, he was disagreeably surprised as the little Gauss stepped forward when the others had scarcely started working, put his slate on the teacher's desk, and said, 'Here it is.' The teacher did not even look at little Gauss's slate, for he felt quite sure that the answer must be wrong, but decided to punish the boy severely for this piece of impudence.   He waited till all the other boys had piled their slates on that of little Gauss, and then he pulled it out and looked at it. What was his surprise as he found on the slate just one number and it was the right one!   What was the number and how did little Gauss find it?"

Of course, we do not know exactly how little Gauss did it and we shall never be able to know.   Yet we are free to imagine something that looks reasonable.   Little Gauss was, after all, just a child, although an exceptionally intelligent and precocious child.   It came to him probably more naturally than to other children of his age to grasp the purpose of a question, to pay attention to the essential point.   He just represented to himself more clearly and distinctly than the other youngsters *what is required*: to find the sum

1
2
3
and so on

.

.

.

20
—

He must have seen the problem differently, more completely, than the others, perhaps with some variations as the successive diagrams *A*, *B*, *C*, *D*, and *E* of Fig. 3.1 indicate. The original statement of the problem emphasizes the beginning of the series of numbers that should be added (*A*). Yet we could also emphasize the end (*B*) or, still better, emphasize the beginning and the end *equally* (*C*). Our attention may attach itself to the two extreme numbers, the very first and the very last, and we may observe some particular relation between them (*D*). Then the idea appears (*E*). Yes, numbers equally removed from the extremes add up all along to the same sum

$$1 + 20 = 2 + 19 = 3 + 18 = \cdots = 10 + 11 = 21$$

and, therefore, the grand total of the whole series is

$$10 \times 21 = 210$$

Did little Gauss really do it this way? I am far from asserting that. I say only that it would be natural to solve the problem in some such way. How did we solve it? Eventually we understood the situation (*E*), we "saw the truth clearly and distinctly," as Descartes would say, we saw a convenient, effortless, well-adapted manner of doing the required sum. How did we reach this final stage? At the outset, we hesitated between two opposite ways of conceiving the problem (*A* and *B*) which we finally

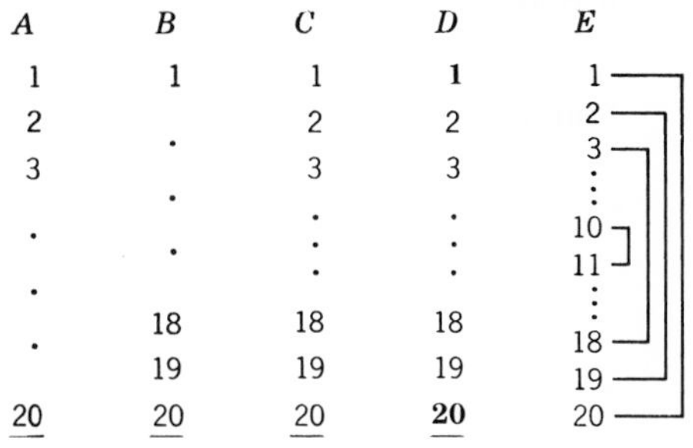

A	*B*	*C*	*D*	*E*
1	1	1	1	1
2	.	2	2	2
3	.	3	3	3
.	.	.	.	.
.	.	.	.	10
.	.	.	.	11
.	18	18	18	.
.	19	19	19	18
20	20	20	**20**	19
				20

Fig. 3.1. Five phases of a discovery.

succeeded in merging into a better balanced conception (*C*). The original antagonism resolved into a happy harmony and the transition (*D*) to the final idea was quite close. Was little Gauss's final idea the same? Did he arrive at it passing through the same stages? Or did he skip some of them? Or did he skip all of them? Did he jump right away at the final conclusion? We cannot answer such questions. Usually a bright idea emerges after a longer or shorter period of hesitation and suspense. This happened in our case, and some such thing may have happened in the mind of little Gauss.

Let us generalize. Starting from the problem just solved and substituting the general positive integer *n* for the particular value 20, we arrive at the problem: *Find the sum S of the first n positive integers.*

Thus we seek the sum

$$S = 1 + 2 + 3 + \cdots + n$$

The idea developed in the foregoing (which might have been that of little Gauss) was to pair off the terms: a term that is at a certain distance from the beginning is paired with another term at the same distance from the end. If we are somewhat familiar with algebraic manipulations, we are easily led to the following modification of this scheme.

We write the sum twice, the second time reversing the original order:

$$S = 1 + \quad 2 \quad + \quad 3 \quad + \cdots + (n-2) + (n-1) + n$$
$$S = n + (n-1) + (n-2) + \cdots + \quad 3 \quad + \quad 2 \quad + 1$$

The terms paired with each other by the foregoing solution appear here conveniently aligned, one written under the other. Adding the two equations we obtain

$$2S = (n+1) + (n+1) + (n+1) + \cdots + (n+1) + (n+1) + (n+1)$$
$$2S = n(n+1)$$
$$S = \frac{n(n+1)}{2}$$

This is the general formula. For $n = 20$ it yields little Gauss's result, which is as it should be.

3.2. Out of the blue

Here is a problem similar to that solved in the foregoing section: *Find the sum of the first n squares.*

Let *S* stand for the required sum (we are no longer bound by the notation of the foregoing section) so that now

$$S = 1 + 4 + 9 + 16 + \cdots + n^2$$

The evaluation of this sum is not too obvious. Human nature prompts us to repeat a procedure that has succeeded before in a similar situation; and so, remembering the foregoing section, we may attempt to write the sum twice, reversing the order the second time:

$$S = 1 + \quad 4 \quad + \quad 9 \quad + \cdots + (n-2)^2 + (n-1)^2 + n^2$$
$$S = n^2 + (n-1)^2 + (n-2)^2 + \cdots + \quad 9 \quad + \quad 4 \quad + 1$$

Yet the addition of these two equations, which was so successful in the foregoing case, leads us nowhere in the present case: our attempt fails, we undertook it with more optimism than understanding, our servile imitation of the chosen pattern was, let us confess, silly. (It was an overdose of mental inertia: our mind persevered in the same course, although this course should have been changed by the influence of circumstances.) Yet even such a misconceived trial need not be quite useless; it may lead us to a more adequate appraisal of the proposed problem: yes, it seems to be more difficult than the problem in the foregoing section.

Well, here is a solution. We start from a particular case of a well-known formula:

$$(n + 1)^3 = n^3 + 3n^2 + 3n + 1$$

from which follows

$$(n + 1)^3 - n^3 = 3n^2 + 3n + 1$$

This is valid for any value of n; write it down successively for $n = 1, 2, 3, \ldots, n$:

$$2^3 \quad - 1^3 = 3 \cdot 1^2 + 3 \cdot 1 + 1$$
$$3^3 \quad - 2^3 = 3 \cdot 2^2 + 3 \cdot 2 + 1$$
$$4^3 \quad - 3^3 = 3 \cdot 3^2 + 3 \cdot 3 + 1$$
$$\cdot \quad \cdot \quad \cdot \quad \cdot \quad \cdot \quad \cdot \quad \cdot \quad \cdot \quad \cdot \quad \cdot$$
$$(n + 1)^3 - n^3 = 3n^2 \quad + 3n \quad + 1$$

What is the obvious thing to do with these n equations? Add them! Thanks to conspicuous cancellations, the left-hand side of the resulting equation will be very simple. On the right-hand side we have to add three columns. The first column brings in S, the desired sum of the squares—that's good! The last column consists of n units—that is easy. The column in the middle brings in the sum of the first n numbers—but we know this sum from the foregoing section. We obtain

$$(n + 1)^3 - 1 = 3S + 3 \frac{n(n + 1)}{2} + n$$

and in this equation everything is known (that is, expressed in terms of n)

except S, and so we can determine S from the equation.   In fact, we find by straightforward algebra

$$2(n^3 + 3n^2 + 3n) = 6S + 3(n^2 + n) + 2n$$

$$S = \frac{2n^3 + 3n^2 + n}{6}$$

or finally

$$S = \frac{n(n + 1)(2n + 1)}{6}$$

How do you like this solution?

I shall be highly pleased with the reader who is displeased with the foregoing solution provided that he gives the right reason for his displeasure. What is wrong with the solution?

The solution is certainly correct.   Moreover, it is efficient, clear, and short.   Remember that the problem appeared difficult—we cannot reasonably expect a much clearer or shorter solution.   There is, as far as I can see, just one valid objection: the solution appears *out of the blue*, pops up from nowhere.   It is like a rabbit pulled out of a hat.   Compare the present solution with that in the foregoing section.   There we could visualize to some extent how the solution was discovered, we could learn a little about the ways of discovery, we could even gather some hope that some day we shall succeed in finding a similar solution by ourselves.   Yet the presentation of the present section gives no hint about the sources of discovery, we are just hit on the head with the initial equation from which everything follows, and there is no indication how we could find this equation by ourselves.   This is disappointing; we came here to learn problem solving—how could we learn it from the solution just presented?[1]

3.3. We cannot leave this unapplied

Yes, we *can* learn something important about problem solving from the foregoing solution.   It is true, the presentation was not enlightening: the source of the invention remained hidden and so the solution appeared as a trick, a cunning device.   Do you wish to know what is behind the trick? Try to *apply that trick* yourself and then you may find out.   The device was so successful that we really cannot afford to leave it unapplied.

Let us start by generalizing.   We bring both problems considered in the foregoing (in sections 3.1 and 3.2) under the same viewpoint by considering the sum of the kth powers of the first n natural numbers

$$S_k = 1^k + 2^k + 3^k + \cdots + n^k$$

[1] Cf. MPR, vol. 2, pp. 146–148, the sections on "*deus ex machina*" and "heuristic justification."

We found in the foregoing section

$$S_2 = \frac{n(n + 1)(2n + 1)}{6}$$

and before that

$$S_1 = \frac{n(n + 1)}{2}$$

to which we may add the obvious, but perhaps not useless, extreme case

$$S_0 = n$$

Starting from the particular cases $k = 0$, 1, and 2 we may raise the general problem: express S_k similarly. Surveying those particular cases, we may even conjecture that S_k can be expressed as a polynomial of degree $k + 1$ in n.

It is natural to try on the general case the trick that served us so well in the case $k = 2$. Yet let us first examine the next particular case $k = 3$. We have to imitate what we have seen in sect. 3.2 on the next higher level—this cannot be very difficult.

In fact, we start by applying the binomial formula with the next higher exponent 4:

$$(n + 1)^4 = n^4 + 4n^3 + 6n^2 + 4n + 1$$

from which follows

$$(n + 1)^4 - n^4 = 4n^3 + 6n^2 + 4n + 1$$

This is valid for any value of n; write it down successively for $n = 1, 2, 3, \ldots, n$:

$$
\begin{aligned}
2^4 \quad &- 1^4 = 4 \cdot 1^3 + 6 \cdot 1^2 + 4 \cdot 1 + 1 \\
3^4 \quad &- 2^4 = 4 \cdot 2^3 + 6 \cdot 2^2 + 4 \cdot 2 + 1 \\
4^4 \quad &- 3^4 = 4 \cdot 3^3 + 6 \cdot 3^2 + 4 \cdot 3 + 1
\end{aligned}
$$

$$\cdot \quad \cdot \quad \cdot \quad \cdot \quad \cdot \quad \cdot \quad \cdot \quad \cdot \quad \cdot \quad \cdot \quad \cdot \quad \cdot$$

$$(n + 1)^4 - n^4 = 4n^3 \quad + 6n^2 \quad + 4n \quad + 1$$

As before, we add these n equations. There are conspicuous cancellations on the left-hand side. On the right-hand side, there are four columns to add, and each column involves a sum of like powers of the first n integers; in fact, each column introduces another particular case of S_k:

$$(n + 1)^4 - 1 = 4S_3 + 6S_2 + 4S_1 + S_0$$

Yet we can already express S_2, S_1, and S_0 in terms of n, see above. Using those expressions, we transform our equation into

$$(n + 1)^4 - 1 = 4S_3 + 6\frac{n(n + 1)(2n + 1)}{6} + 4\frac{n(n + 1)}{2} + n$$

and in this equation everything is expressed in terms of n except S_3. What is needed now to determine S_3 is merely a little straightforward algebra:

$$4S_3 = (n + 1)^4 - (n + 1) - 2n(n + 1) - n(n + 1)(2n + 1)$$
$$= (n + 1)[n^3 + 3n^2 + 3n - 2n - n(2n + 1)]$$
$$S_3 = \left[\frac{n(n + 1)}{2}\right]^2$$

We have arrived at the desired result, and even the route seems instructive: having used the trick a second time, we may foresee a general outline. Remember that dictum of a famous pedagogue: "A method is a device which you use twice."[2]

3.4. Recursion

What was the salient feature of our work in the preceding sect. 3.3? In order to obtain S_3, we went back to the previously determined S_2, S_1, and S_0. This illuminates the "trick" of sect. 3.2 where we obtained S_2 by recurring to the previously determined S_1 and S_0.

In fact, we could use the same scheme to derive S_1 which we obtained in sect. 3.1 by a quite different method. By a most familiar formula

$$(n + 1)^2 = n^2 + 2n + 1$$
$$(n + 1)^2 - n^2 = 2n + 1$$

We list particular cases:

$$2^2 \quad - 1^2 = 2 \cdot 1 + 1$$
$$3^2 \quad - 2^2 = 2 \cdot 2 + 1$$
$$4^2 \quad - 3^2 = 2 \cdot 3 + 1$$

$$\cdot \quad \cdot \quad \cdot \quad \cdot \quad \cdot \quad \cdot \quad \cdot \quad \cdot$$

$$(n + 1)^2 - n^2 = 2n \quad + 1$$

By adding we obtain

$$(n + 1)^2 - 1 = 2S_1 + S_0$$

Of course, $S_0 = n$ and so

$$S_1 = \frac{(n + 1)^2 - 1 - n}{2} = \frac{n(n + 1)}{2}$$

which is the final result of sect. 3.1.

After having worked the scheme in the particular cases $k = 1$, 2, and 3,

---

[2] HSI, The traditional mathematics professor, p. 208.

we apply it without hesitation to the general sum S_k. We now need the binomial formula with the exponent $k + 1$:

$$(n + 1)^{k+1} = n^{k+1} + \binom{k + 1}{1}n^k + \binom{k + 1}{2}n^{k-1} + \cdots + 1$$

$$(n + 1)^{k+1} - n^{k+1} = (k + 1)n^k + \binom{k + 1}{2}n^{k-1} + \cdots + 1$$

We list particular cases:

$$2^{k+1} \quad - 1^{k+1} = (k + 1)1^k + \binom{k + 1}{2}1^{k-1} + \cdots + 1$$

$$3^{k+1} \quad - 2^{k+1} = (k + 1)2^k + \binom{k + 1}{2}2^{k-1} + \cdots + 1$$

$$4^{k+1} \quad - 3^{k+1} = (k + 1)3^k + \binom{k + 1}{2}3^{k-1} + \cdots + 1$$

$$\cdot \quad \cdot \quad \cdot \quad \cdot \quad \cdot \quad \cdot \quad \cdot \quad \cdot \quad \cdot \quad \cdot \quad \cdot \quad \cdot \quad \cdot \quad \cdot \quad \cdot \quad \cdot \quad \cdot$$

$$(n + 1)^{k+1} - n^{k+1} = (k + 1)n^k + \binom{k + 1}{2}n^{k-1} + \cdots + 1$$

By adding we obtain

$$(n + 1)^{k+1} - 1 = (k + 1)S_k + \binom{k + 1}{2}S_{k-1} + \cdots + S_0$$

From this equation we can determine (express in terms of n) S_k provided that we have previously determined S_{k-1}, $S_{k-2}, \ldots$, S_1 and S_0. For example, as we have obtained in the foregoing expressions for S_0, S_1, S_2, and S_3, we could derive an expression for S_4 by straightforward algebra. Having obtained S_4, we could proceed to S_5, and so on.[3]

Thus, by following up the "trick" of sect. 3.2, which appeared "out of the blue," we have arrived at a pattern which deserves to be formulated and remembered with a view to further applications. When we are facing a well-ordered sequence (such as S_0, S_1, S_2, $S_3, \ldots$, $S_k, \ldots$) there is a chance to evaluate the terms of the sequence one at a time. We need two things.

First, the initial term of the sequence should be known somehow (the evaluation of S_0 was obvious).

Second, there should be some relation linking the general term of the sequence to the foregoing terms (S_k is linked to S_0, $S_1, \ldots$, S_{k-1} by the final equation of the present section, foreshadowed by the "trick" of sect. 3.2).

Then we can find the terms one after the other, successively, *recursively*,

[3] This method is due to Pascal; see *Œuvres de Blaise Pascal*, edited by L. Brunschvicg and P. Boutroux, vol. 3, pp. 341–367.

by going back or recurring each time to the foregoing terms. This is the important pattern of *recursion*.

3.5. Abracadabra

The word "abracadabra" means something like "complicated nonsense." We use the word contemptuously today, but there was a time when it was a magic word, engraved on amulets in mysterious forms (like Fig. 3.2 in some respect), and people believed that such an amulet would protect the wearer from disease and bad luck.

In how many ways can you read the word "abracadabra" in Fig. 3.2? It is understood that we begin with the uppermost *A* (the north corner) and read down, passing each time to the next letter (southeast or southwest) till we reach the last *A* (the south corner).

The question is curious. Yet your interest may be really aroused if you notice that there is something familiar behind it. It may remind you of walking or driving in a city. Think of a city that consists of perfectly square blocks, where one-half of the streets run from northwest to southeast and the other streets (or avenues) crossing the former run from northeast to southwest. Reading the magic word of Fig. 3.2 corresponds to a zigzag path in the network of such streets. When you walk along the zigzag path emphasized in Fig. 3.3, you walk ten blocks from the initial *A* to the final *A*. There are several other paths which are ten blocks long between these two endpoints in this network of streets, but there is no path that would be shorter. *Find the number of the different shortest paths in the network between the given endpoints*—this is the general, really

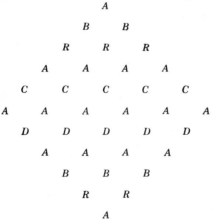

Fig. 3.2. A magic word.

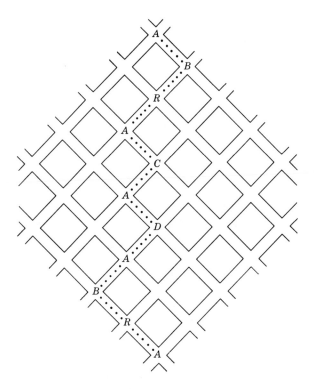

Fig. 3.3. The zigzag path is the shortest.

interesting, problem behind the curious particular problem about the magic word of Fig. 3.2.

A general formulation may have various advantages. It sometimes suggests an approach to the solution, and this happens in our case. *If you cannot solve the proposed problem* about Fig. 3.2 (probably you cannot), *try first to solve some simpler related problem.* At this point the general formulation may help: it suggests trying simpler cases that fall under it. In fact, if the two given corners are close enough to each other in the network (closer than the extreme *A*'s in Fig. 3.3) it is easy to count the different zigzag paths between the two: you can draw each one after the other and survey all of them. Listen to this suggestion and pursue it systematically. Start from the point *A* and go downward. Consider first the points that you can reach by walking one block, then those to which you have to walk two blocks, then those which are three or four or more blocks away. Survey and count for each point the shortest zigzag paths

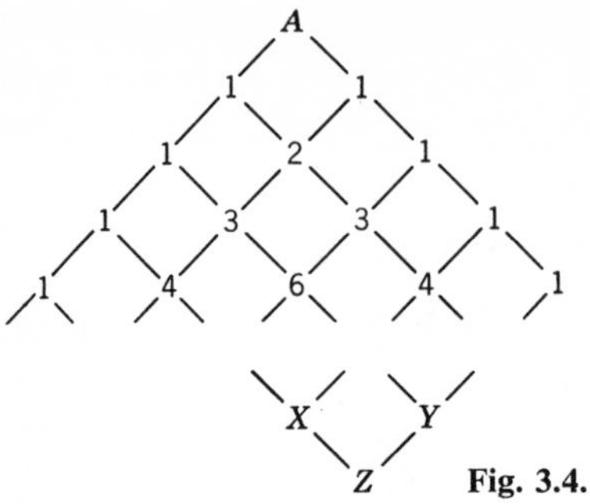

Fig. 3.4. Count the number of shortest zigzag paths.

that connect it with A. In Fig. 3.4 a few numbers so obtained are marked (but you should have obtained these numbers and a few more by yourself— check them at least). Observe these numbers—do you notice something?

If you have enough previous knowledge you may notice many things. Yet even if you have never before seen this array of numbers displayed by Fig. 3.4 you may notice an interesting relation: any number in Fig. 3.4 that is different from 1 is the sum of two other numbers in the array, of its northwest and northeast neighbors. For instance,

$$4 = 1 + 3, \qquad 6 = 3 + 3$$

You may discover this law by observation as a naturalist discovers the laws of his science by observation. Yet, after having discovered it, you should ask yourself: Why is that so? What is the reason?

The reason is simple enough. Consider three corners in your network, the points X, Y, and Z, the relative position of which is shown by Fig. 3.4: X is the northwest neighbor and Y the northeast neighbor of Z. If we wish to reach Z coming from A along a shortest path in the network, we must pass either through X or through Y. Once we have reached X, we can proceed hence to Z in just one way, and the same is true for proceeding from Y to Z. Therefore, the *total number of shortest paths from A to Z* is a sum of two terms: it *equals the number of shortest paths from A to X added to the number of those from A to Y*. This explains fully our observation and proves the general law.

Having clarified this basic point, we can extend the array of numbers in Fig. 3.4 by simple additions till we obtain the larger array in Fig. 3.5, the south corner of which yields the desired answer: we can read the magic word in Fig. 3.2 in exactly 252 different ways.

3.6. The Pascal triangle

By now the reader has probably recognized the numbers and their peculiar configuration which we have examined in the foregoing section. The numbers in Fig. 3.4 are *binomial coefficients* and their triangular arrangement is usually called the *Pascal triangle*. (Pascal himself called it the "arithmetical triangle.") Further lines can be added to the triangle of Fig. 3.4 and, in fact, it can be extended indefinitely. The array in Fig. 3.5 is a square piece cut out of a larger triangle.

Some of the binomial coefficients and their triangular arrangement can be found in the writings of other authors before Pascal's *Traité du triangle arithmétique*. Still, the merits of Pascal in this matter are quite sufficient to justify the use of his name.

(1) We have to introduce a suitable *notation* for the numbers contained in the Pascal triangle; this is a step of major importance. For us each number attached to a point of this triangle has a geometric meaning: it indicates the number of different shortest zigzag paths from the apex of the triangle to that point. Each of these paths passes along the same number of blocks, let us say n blocks. Moreover, all these paths agree in the number of blocks described in the southwesterly direction and in the number of those in the southeasterly direction. Let l and r stand for these numbers, respectively (l to the left and r to the right—of course, downward in both cases). Obviously

$$n = l + r$$

If we give any two of the three numbers n, l, and r, the third is fully determined and so is the point to which they refer. (In fact, l and r are the rectangular coordinates of the point with respect to a system the origin of which is the apex of the Pascal triangle; one of the axes points southwest,

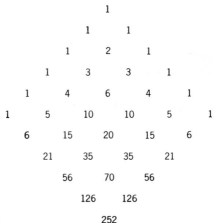

Fig. 3.5. A square from a triangle.

the other southeast.)   For instance, for the last A of the path shown in Fig. 3.3

$$l = 5, \qquad r = 5, \qquad n = 10$$

and for the second B of the same path

$$l = 5, \qquad r = 3, \qquad n = 8$$

We shall denote by $\binom{n}{r}$ (this notation is due to Euler) the number of shortest zigzag paths from the apex of the Pascal triangle to the point specified by n (total number of blocks) and r (blocks to the right downward).   For instance, see Fig. 3.5,

$$\binom{8}{3} = 56, \qquad \binom{10}{5} = 252$$

The symbols for the numbers contained in Fig. 3.4 are assembled in Fig. 3.6.   The symbols with the same number upstairs (the same n) are horizontally aligned (along the nth "base"—the base of a right triangle). The symbols with the same number downstairs (the same r) are obliquely aligned (along the rth "avenue").   The fifth avenue forms one of the sides of the square in Fig. 3.5—the opposite side is formed by the 0th avenue (but you may call it the borderline, or Riverside Drive, if you prefer to do so).   The fourth base is emphasized in Fig. 3.4.

(2) Besides the geometric aspect, the Pascal triangle also has a computational aspect.   All the numbers along the boundary (0th street, 0th avenue, and their common starting point) are equal to 1 (it is obvious that

$$\binom{0}{0}$$

$$\binom{1}{0} \qquad \binom{1}{1}$$

$$\binom{2}{0} \qquad \binom{2}{1} \qquad \binom{2}{2}$$

$$\binom{3}{0} \qquad \binom{3}{1} \qquad \binom{3}{2} \qquad \binom{3}{3}$$

$$\binom{4}{0} \qquad \binom{4}{1} \qquad \binom{4}{2} \qquad \binom{4}{3} \qquad \binom{4}{4}$$

$$\binom{n}{r-1} \qquad \binom{n}{r}$$

$$\binom{n+1}{r}$$

Fig. 3.6. Symbolic Pascal triangle.

there is just one shortest path to these street corners from the starting point). Therefore,

$$\binom{n}{0} = \binom{n}{n} = 1$$

It is appropriate to call this relation the *boundary condition* of the Pascal triangle.

Any number inside the Pascal triangle is situated along a certain horizontal row, or base. We compute a number of the $(n + 1)$th base by going back, or recurring, to two neighboring numbers of the nth base:

$$\binom{n + 1}{r} = \binom{n}{r} + \binom{n}{r - 1}$$

see Fig. 3.6. It is appropriate to call this equation the *recursion formula* of the Pascal triangle.

From the computer's standpoint the numbers $\binom{n}{r}$ are determined (or defined, if you wish) by the recursion formula and the boundary condition of the Pascal triangle.

3.7. Mathematical induction

When we compute a number in the Pascal triangle by using the recursion formula, we have to rely on the previous knowledge of two numbers of the foregoing base. It would be desirable to have a scheme of computation independent of such previous knowledge. There is a well-known formula, which we shall call the *explicit formula* for binomial coefficients, that yields such an independent computation:

$$\binom{n}{r} = \frac{n(n - 1)(n - 2)\cdots(n - r + 1)}{1 \cdot 2 \cdot 3 \cdots r}$$

Pascal's treatise contains the explicit formula (stated in words, not in our modern notation). Pascal does not say how he has discovered it and we shall not speculate too much how he might have discovered it. (Perhaps he just guessed it first—we often find such things by observation and tentative generalization of the observed; see the remark in the solution of ex. 3.39.) Yet Pascal gives a remarkable proof for the explicit formula and we wish to devote our full attention to his method of proof.[4]

We need a preliminary remark. The explicit formula does not apply,

---

[4] Cf. Pascal's *Œuvres l.c.* footnote 3, pp. 455–464, especially pp. 456–457. The following presentation takes advantage of modern notation and modifies less essential details.

as it stands, to the case $r = 0$.   Yet we lay down the rule that, if $r = 0$, it should be interpreted as

$$\binom{n}{0} = 1$$

The explicit formula does apply to the case $r = n$ and yields

$$\binom{n}{n} = \frac{n(n-1)\cdots \quad 2 \cdot 1}{1 \cdot 2 \quad \cdots (n-1)n} = 1$$

which is the correct result.   Therefore, we have to prove the explicit formula only for $0 < r < n$, that is, in the interior of the Pascal triangle where we can use the recursion formula.   Now, we quote Pascal, with unessential modifications some of which will be included in square brackets [ ].

Although this proposition [the explicit formula] contains infinitely many cases I shall give for it a very short proof, supposing two lemmas.

The first lemma asserts that the proposition holds for the first base, which is obvious.   [The explicit formula is valid for $n = 1$, because, in this case, all possible values of r, $r = 0$ and $r = 1$, fall under the preliminary remark.]

The second lemma asserts this: if the proposition happens to be valid for any base [for any value n] it is necessarily valid for the next base [for $n + 1$].

We see hence that the proposition holds necessarily for all values of n. For it is valid for $n = 1$ by virtue of the first lemma; therefore, for $n = 2$ by virtue of the second lemma; therefore, for $n = 3$ by virtue of the same, and so on *ad infinitum*.

And so nothing remains but to prove the second lemma.

In accordance with the statement of the second lemma, we assume that the explicit formula is valid for the nth base, that is, for a certain value of n and all compatible values of r (for $r = 0, 1, 2,\ldots, n$).   In particular, along with

$$\binom{n}{r} = \frac{n(n-1)\cdots(n-r+2)(n-r+1)}{1 \cdot 2 \quad \cdots \quad (r-1) \cdot \quad r}$$

we also have (if $r \geqq 1$)

$$\binom{n}{r-1} = \frac{n(n-1)\cdots(n-r+2)}{1 \cdot 2 \quad \cdots \quad (r-1)}$$

Adding these two equations and using the recursion formula, we derive as a necessary consequence

$$\binom{n+1}{r} = \binom{n}{r} + \binom{n}{r-1} = \frac{n(n-1)\cdots(n-r+2)}{1 \cdot 2 \quad \cdots \quad (r-1)}\left[\frac{n-r+1}{r} + 1\right]$$

$$= \frac{n(n-1)\cdots(n-r+2)}{1 \cdot 2 \quad \cdots \quad (r-1)} \cdot \frac{n+1}{r}$$

$$= \frac{(n+1)n(n-1)\cdots(n-r+2)}{1 \cdot 2 \cdot 3 \quad \cdots \quad r}$$

That is, the validity of the explicit formula for a certain value of n involves its validity for $n + 1$. This is precisely what the second lemma asserts—we have proved it.

The words of Pascal which we have quoted are of historic importance because his proof is the first example of a fundamental pattern of reasoning which is usually called *mathematical induction*.

This pattern of reasoning deserves further study.[5] If carelessly introduced, reasoning by mathematical induction may puzzle the beginner; in fact, it may appear as a devilish trick.

You know, of course, that the devil is dangerous: if you give him the little finger, he takes the whole hand. Yet Pascal's second lemma does exactly this: by admitting the first lemma you give just one finger, the case $n = 1$. Yet then the second lemma also takes your second finger (the case $n = 2$), then the third finger ($n = 3$), then the fourth, and so on, and finally takes all your fingers even if you happen to have infinitely many.

3.8. Discoveries ahead

After the work in the three foregoing sections, we now have three different approaches to the numbers in the Pascal triangle, the binomial coefficients.

(1) *Geometrical approach.* A binomial coefficient is the number of the different shortest zigzag paths between two given corners in a network of streets.

(2) *Computational approach.* The binomial coefficients can be defined by their recursion formula and their boundary condition.

(3) *Explicit formula.* We have proved it, by Pascal's method, in sect. 3.7.

The name of the numbers considered reminds us of another approach.

(4) *Binomial theorem.* For indeterminate (or variable) x and any non-negative integer n we have the identity

$$(1 + x)^n = \binom{n}{0} + \binom{n}{1}x + \binom{n}{2}x^2 + \cdots + \binom{n}{n}x^n$$

For a proof, see ex. 3.1.

There are still other approaches to the numbers in the Pascal triangle which play, in fact, a role in a great many interesting questions and possess a great many interesting properties. "This table of numbers has eminent

---

[5] HSI, Induction and mathematical induction, pp. 114–121; MPR, vol. 1, pp. 108–120.

and admirable properties" wrote Jacob Bernoulli in his *Ars Conjectandi* (Basle 1713; see Second Part, Chapter III, p. 88). "We have just shown that the essence of combinations is concealed in it [see ex. 3.22–3.27] but those who are more intimately acquainted with Geometry know also that capital secrets of all Mathematics are hidden in it." Times have changed and many things hidden in Bernoulli's time are clearly seen today. Still, the reader who wants instructive, and perhaps fascinating, exercise has an excellent opportunity: he has an excellent chance to discover something by observing the numbers in the Pascal triangle and combining his observations with one or the other or several approaches. There are so many possibilities—some of them should be favorable.

By the way, we have broached another subject in the first four sections of the chapter (sum of like powers of the first *n* integers). Moreover, we have encountered two important general patterns (recursion and mathematical induction) which we still should apply to more examples if we wish to understand them thoroughly. And so there are still more prospects ahead.

3.9. Observe, generalize, prove, and prove again

Let us return to our starting point and have another look at it.

(1) We started from the magic word of Fig. 3.2 and Fig. 3.3, or rather from a problem concerning that word. What was the unknown? The number of shortest zigzag paths in that network of streets from the first *A* to the last *A*, that is, from the north corner of the square to its south corner. Such a zigzag path must cross somewhere the horizontal diagonal of the square. There are six possible crossing points (street corners, *A*'s) along the horizontal diagonal. There are, therefore, six different kinds of zigzag paths in our problem—how many paths are there of each kind? We have here a *new problem*.

Let us be specific. Take a definite crossing point on that horizontal diagonal, for instance the third point from the left ($l = 3, r = 2, n = 5$ in the notation of sect. 3.6). A zigzag path crossing this chosen point consists of two sections: the upper section starts from the north corner of the square and ends in the chosen point, the lower section starts from the chosen point and ends in the south corner; see Fig. 3.3. We have found before (see Fig. 3.5) the number of the different upper sections; it is

$$\binom{5}{2} = 10$$

The number of the different lower sections is the same. Now any upper section can be combined with any lower section to form a full path [as suggested by Fig. 3.7(III)]. Therefore, the number of such paths is

$$\binom{5}{2}^2 = 100$$

Of course, the number of zigzag paths crossing the horizontal diagonal at any other given point can be similarly computed. Hence we find a new solution of our original problem: we can read the magic word of Fig. 3.2 in exactly

$$1 + 25 + 100 + 100 + 25 + 1$$

different ways. This sum must agree with the result found at the end of sect. 3.5; in fact, it equals 252.

(2) *Generalization.* One side of the square considered in Fig. 3.3 consists of five blocks. In generalizing (passing from 5 to n) we find that

$$\binom{n}{0}^2 + \binom{n}{1}^2 + \binom{n}{2}^2 + \cdots + \binom{n}{n}^2 = \binom{2n}{n}$$

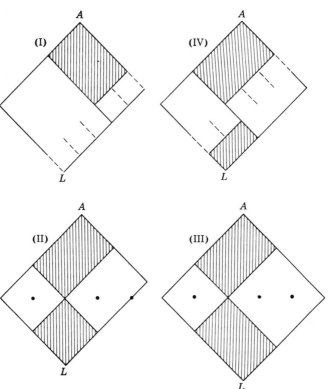

Fig. 3.7. Suggestions.

"The sum of the squares of the numbers in the nth base of the Pascal triangle is equal to the number in the middle of the $2n$th base." Our reasoning under (1) essentially proves this general statement. It is true, we have considered the special case $n = 5$ (we have even considered a special point of the fifth base) but there is no particular virtue (and no misleading peculiarity) in the special case considered. And so our reasoning is generally valid. Yet it may be a useful exercise for the reader to repeat the reasoning with special attention to its generality—he has to say n instead of 5.[6]

(3) *Another approach.* Still, the result is surprising. We would understand it better if we could attain it from another side.

Surveying the various approaches listed in sect. 3.8, we may try to link our result to the binomial formula. There is, in fact, a connection:

$$(1 + x)^{2n} = \cdots + \binom{2n}{n}x^n + \cdots$$

$$= (1 + x)^n(1 + x)^n$$

$$= \left[\binom{n}{0} + \binom{n}{1}x + \binom{n}{2}x^2 + \cdots + \binom{n}{n}x^n\right] \cdot$$

$$\left[\binom{n}{n} + \cdots + \binom{n}{2}x^{n-2} + \binom{n}{1}x^{n-1} + \binom{n}{0}x^n\right]$$

Let us focus the coefficient of x^n. On the right-hand side of the first line the coefficient of x^n is the right-hand side of the general equation given under (2) for which we are seeking a second proof. Now let us turn to the product of the two factors which are displayed on the last two lines; in writing them we made use of the symmetry of the binomial coefficients:

$$\binom{n}{r} = \binom{n}{n-r}$$

Now, in this product, the coefficient of x^n is obviously the left-hand side of the equation under (2) which we are about to prove. And here is the proof: the coefficient of x^n must be the same in both cases since we have here an identity in x.

Examples and Comments on Chapter 3

First Part

The examples and comments of this first part are connected with the first four sections.

[6] We have here a *representative* special case; see MPR, vol. 1, p. 25, ex. 10.

3.1. Prove the binomial theorem stated in sect. 3.8(4) (and used in sect. 3.4). (Use mathematical induction. Which one of the first three approaches mentioned in sect. 3.8 appears the most appropriate for the present purpose?)

3.2. *A particular case equivalent to the general case.* The identity asserted in sect. 3.8(4) and proved in ex. 3.1 follows as a particular case ($a = 1$, $b = x$) from the more general identity

$$(a + b)^n = \binom{n}{0}a^n + \binom{n}{1}a^{n-1}b + \binom{n}{2}a^{n-2}b^2 + \cdots + \binom{n}{n}b^n$$

Show that, conversely, the general identity also follows from that particular case.[7]

3.3. In the first three sections of this chapter we have computed S_k (defined in sect. 3.3) for $k = 1, 2, 3$; the case $k = 0$ is obvious. Comparing these expressions, we may be led to the general theorem: S_k *is a polynomial in n of degree $k + 1$ and the coefficient of its highest term is $1/(k + 1)$.*

This theorem which asserts that

$$S_k = \frac{n^{k+1}}{k+1} + \cdots$$

(where the dots indicate terms of lower degree in n) played an important role in the history of the integral calculus.

Prove the theorem; use mathematical induction.

3.4. We can guess an expression for S_4 by computing numerically the ratio S_4/S_2 for a few small values of n. In fact, for

$$n = 1, \quad 2, \quad 3, \quad 4, \quad 5$$

$$\frac{S_4}{S_2} = 1, \quad \frac{17}{5}, \quad 7, \quad \frac{59}{5}, \quad \frac{89}{5}$$

For the sake of uniformity we write rather

$$\frac{5}{5}, \quad \frac{17}{5}, \quad \frac{35}{5}, \quad \frac{59}{5}, \quad \frac{89}{5}$$

The numerators are close to multiples of 6; in fact, they are

$$6 \cdot 1 - 1, \quad 6 \cdot 3 - 1, \quad 6 \cdot 6 - 1, \quad 6 \cdot 10 - 1, \quad 6 \cdot 15 - 1$$

You should recognize the numbers

$$1, \quad 3, \quad 6, \quad 10, \quad 15$$

If you succeed in constructing an expression for S_4, prove it, independently of sect. 3.4, by mathematical induction.[8]

3.5. Compute S_4, independently of ex. 3.4, by the method indicated in sect. 3.4.

---

[7] Such equivalence of the particular and the general may seem bewildering to the philosopher or to the beginner, but is, in fact, quite usual in mathematics; see MPR, vol. 1, p. 23, ex. 3 and ex. 4.

[8] For broader discussion of a very similar simpler case see MPR, vol. 1, pp. 108–110.

3.6. Show that

$$n = S_0$$
$$n^2 = 2S_1 - S_0$$
$$n^3 = 3S_2 - 3S_1 + S_0$$
$$n^4 = 4S_3 - 6S_2 + 4S_1 - S_0$$

and generally

$$n^k = \binom{k}{1}S_{k-1} - \binom{k}{2}S_{k-2} + \binom{k}{3}S_{k-3} - \cdots + (-1)^{k-1}\binom{k}{k}S_0$$

(This is similar to, but different from, the principal formula of sect. 3.4.)

3.7. Show that

$$S_1 = S_1$$
$$2S_1^2 = 2S_3$$
$$4S_1^3 = 3S_5 + S_3$$
$$8S_1^4 = 4S_7 + 4S_5$$

and generally, for $k = 1, 2, 3, \ldots$,

$$2^{k-1}S_1^k = \binom{k}{1}S_{2k-1} + \binom{k}{3}S_{2k-3} + \binom{k}{5}S_{2k-5} + \cdots$$

The last term on the right-hand side is S_k or kS_{k+1} according as k is odd or even.

(This is similar to ex. 3.6 where, in fact, we could substitute S_0^k for n^k.)

3.8. Show that

$$3S_2 = 3S_2$$
$$6S_2S_1 = 5S_4 + S_2$$
$$12S_2S_1^2 = 7S_6 + 5S_4$$
$$24S_2S_1^3 = 9S_8 + 14S_6 + S_4$$

and generally, for $k = 1, 2, 3, \ldots$,

$$3\cdot 2^{k-1}S_2S_1^{k-1} = \left[\binom{k}{0} + 2\binom{k}{1}\right]S_{2k} + \left[\binom{k}{2} + 2\binom{k}{3}\right]S_{2k-2} + \cdots$$

the last term on the right-hand side is $(k+2)S_{k+1}$ or S_k according as k is odd or even.

3.9. Show that

$$S_3 = S_1^2$$
$$S_5 = S_1^2(4S_1 - 1)/3$$
$$S_7 = S_1^2(6S_1^2 - 4S_1 + 1)/3$$

and generally that S_{2k-1} is a polynomial in $S_1 = n(n+1)/2$, of degree k, divisible by S_1^2 provided that $2k - 1 \geqq 3$. (This generalizes the result of sect. 3.3.)

3.10. Show that

$$S_4 = S_2(6S_1 - 1)/5$$
$$S_6 = S_2(12S_1^2 - 6S_1 + 1)/7$$
$$S_8 = S_2(40S_1^3 - 40S_1^2 + 18S_1 - 3)/15$$

and generally that S_{2k}/S_2 is a polynomial in S_1 of degree $k - 1$. (This generalizes a result encountered in the solution of ex. 3.4.)

3.11. We introduce the notation

$$1^k + 2^k + 3^k + \cdots + n^k = S_k(n)$$

which is more explicit (or specific) than the one introduced in sect. 3.3; k stands for a non-negative integer and n for a positive integer.

We now extend the range of n (but not the range of k): we let $S_k(x)$ denote the polynomial in x of degree $k + 1$ that coincides with $S_k(n)$ for $x = 1, 2, 3, \ldots$; for example,

$$S_3(x) = \frac{x^2(x + 1)^2}{4}$$

Prove that for $k \geq 1$ (not for $k = 0$)

$$S_k(-x - 1) = (-1)^{k-1}S_k(x)$$

3.12. Find $1 + 3 + 5 + \cdots + (2n - 1)$, the sum of the first n odd numbers. (List as many different approaches as you can.)

3.13. Find $1 + 9 + 25 + \cdots + (2n - 1)^2$.

3.14. Find $1 + 27 + 125 + \cdots + (2n - 1)^3$.

3.15. (Continued) Generalize.

3.16. Find $2^2 + 5^2 + 8^2 + \cdots + (3n - 1)^2$

3.17. (Continued) Generalize.

3.18. Find a simple expression for

$$1 \cdot 2 + (1 + 2)3 + (1 + 2 + 3)4 + \cdots + [1 + 2 + \cdots + (n - 1)]n.$$

(Of course, you should try to use suitable points from the foregoing work. What has better prospects to be usable: the results or the method?)

3.19. Consider the $\dfrac{n(n - 1)}{2}$ differences

$$2 - 1,$$
$$3 - 1, \quad 3 - 2$$
$$4 - 1, \quad 4 - 2, \quad 4 - 3$$
$$\cdot \quad \cdot \quad \cdot \quad \cdot \quad \cdot \quad \cdot \quad \cdot \quad \cdot$$
$$n - 1, \quad n - 2, \quad n - 3 \quad , \ldots, \quad n - (n - 1)$$

and compute (*a*) their sum, (*b*) their product, and (*c*) the sum of their squares.

3.20. Define $E_1, E_2, E_3 \ldots$ by the identity

$$x^n - E_1 x^{n-1} + E_2 x^{n-2} - \cdots + (-1)^n E_n$$
$$= (x - 1)(x - 2)(x - 3) \cdots (x - n)$$

Show that

$$E_1 = \frac{n(n + 1)}{2}$$

$$E_2 = \frac{(n - 1)n(n + 1)(3n + 2)}{24}$$

$$E_3 = \frac{(n - 2)(n - 1)n^2(n + 1)^2}{48}$$

$$E_4 = \frac{(n - 3)(n - 2)(n - 1)n(n + 1)(15n^3 + 15n^2 - 10n - 8)}{5760}$$

and show in general that E_k [which should rather be denoted by $E_k(n)$ since it depends on n] is a polynomial of degree $2k$ in n.

[The knowledge of a certain proposition of algebra may be a great help; E_k is the so-called kth elementary symmetric function of the first n integers the sum of the kth powers of which is $S_k = S_k(n)$.   Check $E_k(k) = k!$]

3.21. *Two forms of mathematical induction.*   A typical proposition A that is accessible to proof by mathematical induction has an infinity of cases $A_1, A_2, A_3, \ldots, A_n, \ldots$; in fact, A is equivalent to the simultaneous assertion of $A_1, A_2, A_3, \ldots$.   For instance, if A is the binomial theorem, A_n asserts the validity of the identity.

$$(1 + x)^n = \binom{n}{0} + \binom{n}{1}x + \binom{n}{2}x^2 + \cdots + \binom{n}{n}x^n$$

see ex. 3.1; the binomial theorem asserts, in fact, that this identity holds for $n = 1, 2, 3, 4, \ldots$.

Let us consider three statements about the sequence of propositions $A_1, A_2, A_3, \ldots$:

(I) A_1 is true.
(IIa) A_n implies A_{n+1}.
(IIb) $A_1, A_2, A_3, \ldots A_{n-1}$ and A_n jointly imply A_{n+1}.

Now we can distinguish two procedures.

(*a*) We can conclude from (I) and (IIa) that A_n is true generally, for $n = 1, 2, 3, \ldots$; we drew this conclusion, with Pascal, in sect. 3.7.

(*b*) We can conclude from (I) and (IIb) that A_n is true generally, for $n = 1, 2, 3, \ldots$; we proceeded so in the solution of ex. 3.3.

You may feel that the difference between the procedures (*a*) and (*b*) is more in the form than in the essence.   Could you clarify this feeling and propose a clear argument?

Second Part

3.22. Ten boys went camping together, Bernie, Ricky, Abe, Charlie, Al, Dick, Alex, Bill, Roy, and Artie.   In the evening they divided into two teams of five boys each: one team put up the tent, the other team cooked the supper.

In how many different ways is such a division into two teams possible? (Can a magic word help you?)

3.23. Show that a set of n individuals has $\binom{n}{r}$ different subsets of r individuals. [In more traditional language: the number of *combinations* of n objects taken r at a time is $\binom{n}{r}$.]

3.24. Given n points in the plane in "general position" so that no three points lie on the same straight line. How many straight lines can you draw by joining two given points? How many triangles can you form with vertices chosen among the given points?

3.25. (Continued) Formulate and solve an analogous problem in space.

3.26. Find the number of the diagonals of a convex polygon with n sides.

3.27. Find the number of intersections of the diagonals of a convex polygon of n sides. Consider only points of intersection inside the polygon, and assume that the polygon is "general" so that no three diagonals have a common point.

3.28. A polyhedron has six faces. (We may consider the polyhedron as irregular so that no two of its faces are congruent.) The faces should be painted, one red, two blue, and three brown. In how many different ways can this be done?

3.29. A polyhedron has n faces (no two of which are congruent.) Of these faces, r should be painted red, s sapphire, and t tan; we suppose that $r + s + t = n$. In how many different ways can this be done?

3.30. (Continued) Generalize.

Third Part

In solving some of the following problems, the reader may consider, and choose between, several approaches. (See sect. 3.8; the combinatorial interpretation of the binomial coefficients, cf. ex. 3.23, provides one more access.) The importance of approaching the same problem from several sides was emphasized by Leibnitz. Here is a free translation of one of his remarks: "In comparing two different expressions of the same quantity, you may find an unknown; in comparing two different derivations of the same result, you may find a new method."

3.31. Show in as many ways as you can that

$$\binom{n}{r} = \binom{n}{n-r}$$

3.32. Consider the sum of the numbers along a base of the Pascal triangle:

$$
\begin{aligned}
1 &= 1 \\
1 + 1 &= 2 \\
1 + 2 + 1 &= 4 \\
1 + 3 + 3 + 1 &= 8
\end{aligned}
$$

These facts seem to suggest a general theorem. Can you guess it? Having guessed it, can you prove it? Having proved it, can you devise another proof?

3.33. Observe

$$
\begin{aligned}
1 - 1 &= 0 \\
1 - 2 + 1 &= 0 \\
1 - 3 + 3 - 1 &= 0 \\
1 - 4 + 6 - 4 + 1 &= 0
\end{aligned}
$$

generalize, prove, and prove again.

3.34. Consider the sum of the first six numbers along the third avenue of the Pascal triangle:

$$1 + 4 + 10 + 20 + 35 + 56 = 126$$

Locate this sum in the Pascal triangle, try to observe analogous facts, generalize, prove, and prove again.

3.35. Add the thirty-six numbers displayed in Fig. 3.5, try to locate their sum in the Pascal triangle, formulate a general theorem, and prove it. (Adding so many numbers is a boring task—in doing it cleverly, you may easily catch the essential idea.)

3.36. Try to recognize and locate in the Pascal triangle the numbers involved in the following relation:

$$1 \cdot 1 + 5 \cdot 4 + 10 \cdot 6 + 10 \cdot 4 + 5 \cdot 1 = 126$$

Observe (or remember) analogous cases, generalize, prove, prove again.

3.37. Try to recognize and locate in the Pascal triangle the numbers involved in the following relation:

$$6 \cdot 1 + 5 \cdot 3 + 4 \cdot 6 + 3 \cdot 10 + 2 \cdot 15 + 1 \cdot 21 = 126$$

Observe (or remember) analogous cases, generalize, prove, prove again.

3.38. Fig. 3.8 shows the first four from an infinite sequence of figures each of which is an assemblage of equal circles into an equilateral triangular shape. Any circle that is not on the rim of the assemblage touches six surrounding circles. In the nth figure there are n circles aligned along each side of the triangular assemblage and the total number of circles in this nth figure is termed the nth *triangular number*. Express the nth triangular number in terms of n and locate it in the Pascal triangle.

3.39. Replace in Fig. 3.8 each circle by a sphere (a marble) of which the

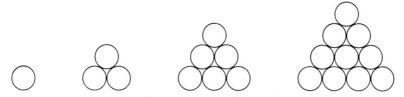

Fig. 3.8. The first four triangular numbers.

circle forms the equator.   Fix 10 marbles arranged as in Fig. 3.8 on a horizontal plane, place 6 marbles on top (they fit neatly into the interstices) as a second
layer, add 3 marbles on top of these as a third layer and place finally 1 marble
on the very top.   This configuration of

$$1 + 3 + 6 + 10 = 20$$

marbles is so related to a regular tetrahedron as each of the assemblages of
circles shown by Fig. 3.8 is related to a certain equilateral triangle: 20 is the
fourth *pyramidal number*.   Express the nth pyramidal number in terms of n
and locate it in the Pascal triangle.

3.40.   You can build a pyramidal pile of marbles in another manner: begin
with a layer of n^2 marbles, arranged in a square as in Fig. 3.9, place on top of
it a second layer of $(n - 1)^2$ marbles, then $(n - 2)^2$ marbles, and so on, and
finally just one marble on the very top.   How many marbles does the pile
contain?

3.41.   Interpret the product

$$\binom{n_1}{r_1}\binom{n_2}{r_2}\binom{n_3}{r_3}\cdots\binom{n_h}{r_h}$$

as the number of a certain set of zigzag paths in a network of streets.

3.42.   All the shortest zigzag paths from the apex of the Pascal triangle to the
point specified by n (the total number of blocks) and r (blocks to the right
downward) have a point in common with the line of symmetry of the Pascal
triangle (from the first A to the last A in Fig. 3.3) namely their common initial

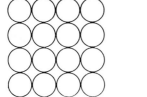
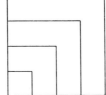
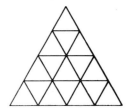

Fig. 3.9. The fourth square number.

point, the apex.   In this set of paths, consider the subset of such paths as have no further point in common with the line of symmetry and find their number N.

In order to realize the meaning of our problem, consider easy particular cases: for

$$r = 0, \quad n, \quad n/2 \ (n \text{ even})$$
$$N = 1, \quad 1, \quad 0$$

Solution.   It will suffice to consider the case $r > n/2$; that is, the common lower endpoint of our zigzag paths lies in the right-hand half of the plane bisected by the line of symmetry.   There are $\binom{n}{r}$ paths in the full set which we divide into three nonoverlapping subsets.

(1) The subset defined above of which we have to find the number of members, N.   A path of the set that does *not* belong to this subset has, besides A, another point on the line of symmetry.

(2) Paths beginning with a block to the left downward; such a path must cross the line of symmetry somewhere since its endpoint lies in the other half plane.   The number of paths in this subset is obviously $\binom{n-1}{r}$.

(3) Such paths as belong neither to (1) nor to (2); they begin with a block to the right downward but subsequently attain somewhere the line of symmetry.

Show that there are just as many paths in subset (2) as in subset (3) (Fig. 3.10 hints the decisive idea of a one-one correspondence between these subsets) and

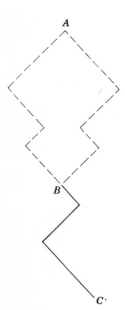

A

B

$C\cdot$ **Fig. 3.10.** The decisive idea.

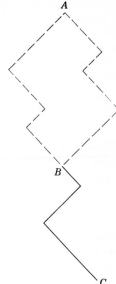

Fig. 3.11. A modification of the decisive idea.

derive hence that

$$N = \frac{|2r - n|}{n} \binom{n}{r}$$

3.43. (Continued) The number of all shortest zigzag paths from the apex to the nth base, that have only the initial point in common with the line of symmetry, is $\binom{2m}{m}$ if $n = 2m$ is even and $2\binom{2m}{m}$ if $n = 2m + 1$ is odd.

3.44. *Trinomial coefficients.* Fig. 3.12 shows a fragment of an infinite triangular array of numbers defined by two conditions.

(1) *Boundary condition.* Each horizontal line or "base" (this term has been similarly used in sect. 3.6) begins with 0, 1 and ends with 1, 0. (The nth base consists of $2n + 3$ numbers and so the boundary condition leaves undefined $2n - 1$ numbers of the nth base, for $n = 1, 2, 3, \ldots$.)

				0	1	0						
			0	1	1	1	0					
		0	1	2	3	2	1	0				
	0	1	3	6	7	6	3	1	0			
0	1	4	10	16	19	16	10	4	1	0		
0	1	5	15	30	45	51	45	30	15	5	1	0

Fig. 3.12. Trinomial coefficients.

(2) *Recursion formula.* Any number of the $(n + 1)$th base left undefined by (1) is computed as the sum of three numbers of the nth base: of its northwestern, northern, and northeastern neighbors. (For instance, $45 = 10 + 16 + 19$.)

Compute the numbers of the seventh base. (They are, with three exceptions, divisible by 7.)

3.45. (Continued) Show that the numbers of the nth base, beginning and ending with 1, are the coefficients in the expansion of $(1 + x + x^2)^n$ in powers of x. (Hence the name "trinomial coefficient.")

3.46. (Continued) Explain the symmetry of Fig. 3.12 with respect to its middle vertical line.

3.47. (Continued) Observe that

$$
\begin{aligned}
1 + 1 + 1 &= 3 \\
1 + 2 + 3 + 2 + 1 &= 9 \\
1 + 3 + 6 + 7 + 6 + 3 + 1 &= 27
\end{aligned}
$$

generalize and prove.

3.48. (Continued) Observe that

$$
\begin{aligned}
1 - 1 + 1 &= 1 \\
1 - 2 + 3 - 2 + 1 &= 1 \\
1 - 3 + 6 - 7 + 6 - 3 + 1 &= 1
\end{aligned}
$$

generalize and prove.

3.49. (Continued) Observe that the value of the sum

$$1^2 + 2^2 + 3^2 + 2^2 + 1^2 = 19$$

is a trinomial coefficient, generalize, and prove.

3.50. (Continued) Find lines in Fig. 3.12 agreeing with lines in the Pascal triangle.

3.51. *Leibnitz's Harmonic Triangle.* Fig. 3.13 shows a section of this little known but remarkable arrangement of numbers. It has properties which are so to say "analogous by contrast" to those of the Pascal triangle. That triangle contains integers, this one (as far as visible) the reciprocals of integers. In Pascal's triangle, each number is the sum of its northwestern and northeastern neighbors. In Leibnitz's triangle, each number is the sum of its southwestern and southeastern neighbors; for instance

$$\frac{1}{2} = \frac{1}{3} + \frac{1}{6}, \qquad \frac{1}{3} = \frac{1}{4} + \frac{1}{12}, \qquad \frac{1}{6} = \frac{1}{12} + \frac{1}{12}$$

This is the *recursion formula* of the Leibnitz triangle. This triangle has also a *boundary condition*: the numbers along the northwest borderline (the "0th avenue") are the reciprocals of the successive integers, $1/1$, $1/2$, $1/3$,.... (The boundary condition of the Pascal triangle is of a different nature: values are prescribed along the whole boundary, 0th avenue *and* 0th street.)

$$\frac{1}{1}$$

$$\frac{1}{2} \qquad \frac{1}{2}$$

$$\frac{1}{3} \qquad \frac{1}{6} \qquad \frac{1}{3}$$

$$\frac{1}{4} \qquad \frac{1}{12} \qquad \frac{1}{12} \qquad \frac{1}{4}$$

$$\frac{1}{5} \qquad \frac{1}{20} \qquad \frac{1}{30} \qquad \frac{1}{20}$$

$$\frac{1}{6} \qquad \frac{1}{30} \qquad \frac{1}{60} \qquad \frac{1}{60}$$

$$\frac{1}{7} \qquad \frac{1}{42} \qquad \frac{1}{105} \qquad \frac{1}{140}$$

$$\frac{1}{8}$$

Fig. 3.13. A fragment of Leibnitz's Harmonic Triangle.

Starting from the given boundary values, we obtain the others by addition in the case of the Pascal triangle, but by subtraction in the present case; in Fig. 3.13 there are some gaps that can be filled immediately with the help of the recursion formula, for instance

$$\frac{1}{4} - \frac{1}{20} = \frac{1}{5} \quad \text{and} \quad \frac{1}{7} - \frac{1}{8} = \frac{1}{56}$$

Using the boundary condition and the recursion formula, extend the table of Fig. 3.13 to the eighth base inclusively.

3.52. *Pascal and Leibnitz.* Try to recognize a connection between corresponding numbers of the two triangles and, having recognized it, prove it.

†**3.53.** Prove[9]

$$\frac{1}{1} = \frac{1}{2} + \frac{1}{6} + \frac{1}{12} + \frac{1}{20} + \frac{1}{30} + \cdots$$

$$\frac{1}{2} = \frac{1}{3} + \frac{1}{12} + \frac{1}{30} + \frac{1}{60} + \frac{1}{105} + \cdots$$

$$\frac{1}{3} = \frac{1}{4} + \frac{1}{20} + \frac{1}{60} + \frac{1}{140} + \frac{1}{280} + \cdots$$

. . . .

(Locate these numbers in the harmonic triangle.)

[9] To accommodate readers who have not had the opportunity to acquire precise ideas about infinite series (limits, convergence,...) the solutions to this and similar problems given on the next pages will not insist on precise details. Yet such details (they are easy in most cases) should be supplied by readers with more complete knowledge.

†**3.54.** Find the sum

$$\frac{1}{12} + \frac{1}{30} + \frac{1}{60} + \frac{1}{105} + \cdots$$

and generalize. (Do you know an analogous problem?)

†**3.55.** Find the sum of the series

$$\frac{1}{1\cdot2} + \frac{1}{2\cdot3} + \frac{1}{3\cdot4} + \frac{1}{4\cdot5} + \cdots$$

$$\frac{1}{1\cdot2\cdot3} + \frac{1}{2\cdot3\cdot4} + \frac{1}{3\cdot4\cdot5} + \frac{1}{4\cdot5\cdot6} + \cdots$$

. . . .

$$\frac{1}{1\cdot2\ldots(r-1)r} + \frac{1}{2\cdot3\ldots r(r+1)} + \frac{1}{3\cdot4\ldots(r+1)(r+2)} + \cdots$$

Fourth Part

Several problems in this part are connected with ex. 3.61, and some others with ex. 3.70.

†**3.56.** *Power series.* The decimal fraction 3.14159...of the number π is, in fact, an "infinite series"

$$3 + 1\left(\frac{1}{10}\right) + 4\left(\frac{1}{10}\right)^2 + 1\left(\frac{1}{10}\right)^3 + 5\left(\frac{1}{10}\right)^4 + 9\left(\frac{1}{10}\right)^5 + \cdots$$

By substituting for $\frac{1}{10}$ the *variable x*, and for the successive digits

$$3, \quad 1, \quad 4, \quad 1, \quad 5, \quad 9, \quad \cdots$$

the *constant coefficients*

$$a_0, \quad a_1, \quad a_2, \quad a_3, \quad a_4, \quad a_5, \quad \cdots$$

respectively, we obtain a *power series*

(1) $a_0 + a_1x + a_2x^2 + a_3x^3 + \cdots$

We are not prepared to consider here the convergence of power series and related important matters, only formal operations with such series (see the footnote appended to ex. 3.53). Multiplication of the proposed power series by a constant c yields

$$ca_0 + ca_1x + ca_2x^2 + ca_3x^3 + \cdots$$

Addition of the proposed power series (1) to another

(2) $b_0 + b_1x + b_2x^2 + b_3x^3 + \cdots$

yields

$$(a_0 + b_0) + (a_1 + b_1)x + (a_2 + b_2)x^2 + (a_3 + b_3)x^3 + \cdots$$

and the product of our two power series (1) and (2) is

$$a_0b_0 + (a_0b_1 + a_1b_0)x + (a_2b_0 + a_1b_1 + a_0b_2)x^2 + \cdots$$

Our two power series (1) and (2) are equal if, and only if,

$$a_0 = b_0, \quad a_1 = b_1, \quad a_2 = b_2, \quad \ldots, \quad a_n = b_n, \quad \ldots$$

We conceive a polynomial as a power series in which infinitely many coefficients, in fact, all coefficients beyond a certain one, vanish. For instance, the polynomial $3x - x^3$ has to be considered as the particular case of our power series (1) in which

$$a_0 = 0, \quad a_1 = 3, \quad a_2 = 0, \quad a_3 = -1 \quad \text{and} \quad a_n = 0 \quad \text{for} \quad n \geqq 4$$

Convince yourself that the foregoing rules are valid for polynomials.

†**3.57.** Compute the product

$$(1 - x)(1 + x + x^2 + \cdots + x^n + \cdots)$$

†**3.58.** Find the coefficient of x^n in the product

$$(a_0 + a_1 x + a_2 x^2 + \cdots + a_n x^n + \cdots)(1 + x + x^2 + \cdots + x^n + \cdots)$$

†**3.59.** A gap in the solution of ex. 3.57 may suggest the consideration of the series

$$
\begin{aligned}
&1 + x + x^2 + x^3 + \cdots \\
&1 + 2x + 3x^2 + 4x^3 + \cdots \\
&1 + 3x + 6x^2 + 10x^3 + \cdots \\
&1 + 4x + 10x^2 + 20x^3 + \cdots
\end{aligned}
$$

Do you know the sum of one of these series? Could you find the sum of the others?

†**3.60.** Give another proof for the result of ex. 3.37.

†**3.61.** *The binomial theorem for fractional and negative exponents.* In a letter of October 24, 1676, addressed to the Secretary of the Royal Society, Newton described how he discovered the (general) binomial theorem; he wrote this letter to answer an inquiry of Leibnitz about his (Newton's) method of discovery.[10] Newton considered the areas under certain curves; he was strongly influenced by ideas of Wallis about interpolation; and eventually he arrived at a conjecture: *The expansion*

$$(1 + x)^a = 1 + \frac{a}{1}x + \frac{a(a - 1)}{1 \cdot 2} x^2 + \frac{a(a - 1)(a - 2)}{1 \cdot 2 \cdot 3} x^3 + \cdots$$

is valid not only for positive integral values of the exponent a, but also for fractional and negative values, in fact, for all numerical values of a.[11]

Newton did not produce a formal proof for his conjecture; rather he relied on examples and analogy. He investigated the question, we may say, as a physicist, "experimentally" or "inductively." In order to understand his viewpoint, we shall try to retrace some of the steps that convinced him of the

[10] Cf. J. R. Newman, *The World of Mathematics*, vol. 1, pp. 519–524.

[11] We know today that some restriction concerning x is necessary but we disregard it here. Such neglect agrees with Newton's standpoint in whose time the convergence of a series was not explicitly defined, and it agrees also with the standpoint of footnote 9.

soundness of his conjecture, which we shall call, for the sake of brevity, the "conjecture N."

If a is a non-negative integer, the coefficient of x^{a+1} vanishes on the right-hand side of the proposed series, and all the following coefficients vanish (thanks to the presence of a factor 0 in the numerator): the series terminates. If, however, a has a value not contained in the sequence 0, 1, 2, 3, ..., the series does not terminate but goes to infinity. For example, for $a = \frac{1}{2}$ the expansion under scrutiny turns out to be

$$(1 + x)^{\frac{1}{2}} = 1 + \frac{\frac{1}{2}}{1}x + \frac{\frac{1}{2}(-\frac{1}{2})}{1 \cdot 2} x^2 + \frac{\frac{1}{2}(-\frac{1}{2})(-\frac{3}{2})}{1 \cdot 2 \cdot 3} x^3 + \cdots$$

$$= 1 + \frac{x}{2} - \frac{x^2}{8} + \frac{x^3}{16} - \frac{5x^4}{128} + \cdots$$

Newton did not appear to be disturbed by an infinity of nonvanishing terms. He knew very well the analogy, which he mentions elsewhere, between power series and decimal fractions; see ex. 3.56. Now, some decimal fractions terminate (as that for $\frac{1}{2}$ or $\frac{3}{8}$) and others do not (as that for $\frac{1}{3}$ or $\frac{7}{11}$, for instance).

Is the above series for $(1 + x)^{\frac{1}{2}}$ valid? To examine this question, Newton multiplied the series by itself; the result *should* be

$$(1 + x)^{\frac{1}{2}}(1 + x)^{\frac{1}{2}} = 1 + x$$

To check this, compute the coefficients of x, x^2, x^3, and x^4 in the product series (ex. 3.56).

†**3.62.** Compute the coefficients of x, x^2, x^3, and x^4 in the square of the series

$$1 + \frac{x}{3} - \frac{x^2}{9} + \frac{5x^3}{81} - \frac{10x^4}{243} + \cdots$$

which is the expansion of $(1 + x)^{\frac{1}{3}}$ according to conjecture N. The result should be the expansion, according to the same conjecture, of $(1 + x)^{\frac{2}{3}}$. Check it!

†**3.63.** (Continued) Compute the coefficients of x, x^2, x^3, and x^4 in the cube of the given series. Predict the result and check your prediction.

†**3.64.** Expand $(1 + x)^{-1}$ according to conjecture N. Any comment?

3.65. *Extending the range.* In sect. 3.6, we have defined the symbol $\binom{n}{r}$

for non-negative integers n and r subject to the inequality $r \leqq n$. We now extend the range of n (but not that of r, cf. ex. 3.11): we set

$$\binom{x}{0} = 1, \qquad \binom{x}{r} = \frac{x(x - 1)(x - 2) \cdots (x - r + 1)}{1 \cdot 2 \cdot 3 \cdots r}$$

for $r = 1, 2, 3, \cdots$ and an arbitrary number x. This definition implies:

(I) $\binom{x}{r}$ is a polynomial in x, of degree r, for $r = 0, 1, 2, 3, \cdots$.

(II) $\binom{x}{r} = (-1)^r \binom{r - 1 - x}{r}$

(III) If n and r are non-negative integers and $r > n$, $\binom{n}{r} = 0$.

(IV) Conjecture N can be written in the form

$$(1 + x)^a = \binom{a}{0} + \binom{a}{1}x + \binom{a}{2}x^2 + \cdots + \binom{a}{n}x^n + \cdots$$

(I), (III), and (IV) are obvious; prove (II).

†**3.66.** Generalize ex. 3.64: examine whether the full result of ex. 3.59 agrees with the conjecture N.

†**3.67.** Apply a device which we have already used three times (sect. 3.9, ex. 3.36, and ex. 3.60) once more: Assuming the conjecture N, compute in two different ways the coefficient of x^r in the expansion of

$$(1 + x)^a(1 + x)^b = (1 + x)^{a+b}$$

†**3.68.** Try to assess the result of ex. 3.67: is it proved? Is some part of it proved? Are there other means to prove it? If we took it for granted, could we prove conjecture N? Or could we prove some part of conjecture N?

†**3.69.** Try to recognize the coefficients of the expansion

$$(1 - 4x)^{-\frac{1}{2}} = 1 + 2x + 6x^2 + 20x^3 + \cdots$$

and express the general term in a familiar form (which should render obvious the fact that the coefficients are integers).

†**3.70.** *The method of undetermined coefficients.* Expand the ratio of two given power series in a power series.

We have to expand in powers of the variable x the ratio

$$\frac{b_0 + b_1x + b_2x^2 + \cdots + b_nx^n + \cdots}{a_0 + a_1x + a_2x^2 + \cdots + a_nx^n + \cdots}$$

where the coefficients $a_0, a_1, a_2, \ldots, b_0, b_1, b_2, \ldots$ are given numbers; we assume that $a_0 \neq 0$. (This assumption, which we have not mentioned in the first short statement of the problem, is essential.)

We are required to exhibit the given ratio in the form

$$\frac{b_0 + b_1x + b_2x^2 + \cdots}{a_0 + a_1x + a_2x^2 + \cdots} = u_0 + u_1x + u_2x^2 + \cdots$$

The coefficients $u_0, u_1, u_2, \ldots, u_n, \ldots$ are not yet determined at this moment when we are just introducing them (hence the name of the method which we are about to apply), yet we hope to determine them eventually; in fact, to determine them is precisely the task imposed upon us by the problem; the coefficients $u_0, u_1, u_2, \ldots$ are the unknowns of our problem (which has, as we now see, an infinity of unknowns).

We rewrite the relation between the three power series (two are given, one we are required to find) in the form

$$(a_0 + a_1x + a_2x^2 + \cdots)(u_0 + u_1x + u_2x^2 + \cdots) = b_0 + b_1x + b_2x^2 + \cdots$$

and now we can see the situation in a more familiar light (ex. 3.56): by equating the coefficients of like powers of x on opposite sides, we obtain a system of equations

$$
\begin{aligned}
a_0 u_0 &= b_0 \\
a_1 u_0 + a_0 u_1 &= b_1 \\
a_2 u_0 + a_1 u_1 + a_0 u_2 &= b_2 \\
a_3 u_0 + a_2 u_1 + a_1 u_2 + a_0 u_3 &= b_3
\end{aligned}
$$

$$\cdot \quad \cdot \quad \cdot \quad \cdot \quad \cdot \quad \cdot \quad \cdot \quad \cdot \quad \cdot \quad \cdot$$

This system of equations presents a familiar pattern: it is recursive, that is, it can be solved by recursion. We obtain the initial unknown u_0 from the initial equation and, having obtained u_0, u_1, ..., u_{n-2} and u_{n-1}, we find the next unknown u_n from the next equation not formerly used.

Express u_0, u_1, u_2, and u_3 in terms of a_0, a_1, a_2, a_3, b_0, b_1, b_2, and b_3.

(The foregoing solution can advantageously serve as a pattern. Note the typical steps:

we introduce the unknowns as the coefficients of a power series;
we derive a system of equations by comparing coefficients of like powers on opposite sides of a relation between power series;
we compute the unknowns recursively.

These steps characterize the pattern, or method, of "undetermined coefficients" which yields some of the most remarkable and most useful systems of equations that can be solved by recursion.)

†3.71. We consider the product of powers

$$a_i{}^{\alpha_i} a_j{}^{\alpha_j} a_k{}^{\alpha_k} b_l{}^{\beta_l} b_m{}^{\beta_m}$$

of which, by definition,

$\alpha_i + \alpha_j + \alpha_k + \beta_l + \beta_m$ is the *degree*,
$\alpha_i + \alpha_j + \alpha_k$ the *degree in the a's*.
$\beta_l + \beta_m$ the *degree in the b's*
$i\alpha_i + j\alpha_j + k\alpha_k + l\beta_l + m\beta_m$ the *weight*.

Of course, the terms just defined are applicable if any number of a's and b's are involved, and not just three of the one kind and two of the other.

Observe the expressions for u_0, u_1, u_2, and u_3 you found in answering ex. 3.70 and explain the regularities observed.

†3.72. Expand the ratio

$$\frac{b_0 + b_1 x + b_2 x^2 + \cdots + b_n x^n + \cdots}{1 + x + x^2 + \cdots + x^n + \cdots}$$

(The result is simple—can you use it?)

†3.73. Expand the ratio

$$\frac{b_0 + b_1 x + b_2 x^2 + \cdots + b_n x^n + \cdots}{1 - x}$$

(The result is simple—can you use it?)

†3.74. Expand the ratio

$$\frac{1 + \dfrac{x}{3} + \dfrac{x^2}{15} + \dfrac{x^3}{105} + \cdots + \dfrac{x^n}{3 \cdot 5 \cdot 7 \cdots (2n + 1)} + \cdots}{1 + \dfrac{x}{2} + \dfrac{x^2}{8} + \dfrac{x^3}{48} + \cdots + \dfrac{x^n}{2 \cdot 4 \cdot 6 \cdots 2n} + \cdots}$$

(Compute a few terms and try to guess the general term.)

†3.75. *Inversion of a power series.* Being given the power series of a function, find the power series of the inverse function.

In other words: being given the expansion of x in powers of y, expand y in powers of x.

More precisely: being given

$$x = a_1 y + a_2 y^2 + \cdots + a_n y^n + \cdots$$

assume that $a_1 \neq 0$ and find the expansion

$$y = u_1 x + u_2 x^2 + \cdots + u_n x^n + \cdots$$

We follow the pattern of ex. 3.70. In the given expansion of x in powers of y, we substitute for y its (desired) power series:

$$\begin{aligned}
x = {} & a_1(u_1 x + u_2 x^2 + u_3 x^3 + \cdots) \\
& + a_2(u_1{}^2 x^2 + 2u_1 u_2 x^3 + \cdots) \\
& + a_3(u_1{}^3 x^3 + \cdots) \\
& + \quad . \quad . \quad .
\end{aligned}$$

In equating the coefficients of like powers of x on opposite sides of this relation, we obtain a system of equations for $u_1, u_2, u_3, \ldots$:

$$\begin{aligned}
1 &= a_1 u_1 \\
0 &= a_1 u_2 + a_2 u_1{}^2 \\
0 &= a_1 u_3 + 2a_2 u_1 u_2 + a_3 u_1{}^3 \\
& \quad . \quad . \quad . \quad . \quad . \quad . \quad . \quad . \quad .
\end{aligned}$$

and the system so obtained is recursive (although not linear.)

Express u_1, u_2, u_3, u_4, and u_5 in terms of a_1, a_2, a_3, a_4, and a_5.

†3.76. Examine the degree and weight of the expressions you found in answering ex. 3.75.

†3.77. Being given that

$$x = y + y^2 + y^3 + \cdots + y^n + \cdots$$

expand y in powers of x.

(The result is simple—can you use it?)

†3.78. Being given that

$$4x = 2y - 3y^2 + 4y^3 - 5y^4 + \cdots$$

expand y in powers of x. (Try to guess the form of the general term and, having guessed it, try to explain it.)

†**3.79.** Being given that

$$x = y + ay^2$$

expand y in powers of x. (The result can be used to clarify a detail of the general situation considered in ex. 3.75.)

†**3.80.** Being given that

$$x = y + \frac{y^2}{2} + \frac{y^3}{6} + \frac{y^4}{24} + \cdots + \frac{y^n}{n!} + \cdots$$

expand y in powers of x.

†**3.81.** *Differential equations.* Expand in powers of x the function y that satisfies the *differential equation*

$$\frac{dy}{dx} = x^2 + y^2$$

and the *initial condition*

$$y = 1 \quad \text{for} \quad x = 0.$$

Following the pattern of ex. 3.70, we set

$$y = u_0 + u_1 x + u_2 x^2 + u_3 x^3 + \cdots$$

with coefficients $u_0, u_1, u_2, \ldots$ which we have still to determine. The differential equation requires

$$u_1 + 2u_2 x + 3u_3 x^2 + 4u_4 x^3 + \cdots$$
$$= u_0^2 + 2u_0 u_1 x + (2u_0 u_2 + u_1^2 + 1)x^2 + \cdots$$

Comparing coefficients of like powers on opposite sides of this relation, we obtain the system of equations.

$$u_1 = u_0^2$$
$$2u_2 = 2u_0 u_1$$
$$3u_3 = 2u_0 u_2 + u_1^2 + 1$$
$$4u_4 = 2u_0 u_3 + 2u_1 u_2$$

$$\cdot \quad \cdot \quad \cdot \quad \cdot \quad \cdot \quad \cdot \quad \cdot$$

From this system we can find $u_1, u_2, u_3, \ldots$, recursively, since the initial condition yields

$$u_0 = 1$$

Compute numerically $u_1, u_2, u_3,$ and u_4.

(The solution of differential equations by the method of undetermined coefficients, which is exemplified by our problem, is of great importance both in theory and in practice.)

†**3.82.** (Continued) Show that $u_n > 1$ for $n \geq 3$.

†**3.83.** Expand in powers of x the function y that satisfies the differential equation

$$\frac{d^2 y}{dx^2} = -y$$

and the initial conditions

$$y = 1, \quad \frac{dy}{dx} = 0 \quad \text{for} \quad x = 0$$

†**3.84.** Find the coefficient of x^{100} in the power series expansion of the function

$$(1 - x)^{-1}(1 - x^5)^{-1}(1 - x^{10})^{-1}(1 - x^{25})^{-1}(1 - x^{50})^{-1}$$

There is little doubt that, in order to solve the proposed problem, we have to generalize it and look for ways and means to compute the general coefficient (that of x^n) in the expansion under consideration. It is also advisable to examine the easier analogous problems implied by the proposed problem. Some meditation on these lines may eventually suggest a plan: introduce *several* power series with "undetermined" coefficients. We set

$$(1 - x)^{-1} = A_0 + A_1 x + A_2 x^2 + A_3 x^3 + A_4 x^4 + \cdots$$
$$(1 - x)^{-1}(1 - x^5)^{-1} = B_0 + B_1 x + B_2 x^2 + B_3 x^3 + \cdots$$
$$(1 - x)^{-1}(1 - x^5)^{-1}(1 - x^{10})^{-1} = C_0 + C_1 x + C_2 x^2 + \cdots$$
$$(1 - x)^{-1}(1 - x^5)^{-1}(1 - x^{10})^{-1}(1 - x^{25})^{-1} = D_0 + D_1 x + \cdots$$

and finally

$$(1 - x)^{-1}(1 - x^5)^{-1}(1 - x^{10})^{-1}(1 - x^{25})^{-1}(1 - x^{50})^{-1}$$
$$= E_0 + E_1 x + E_2 x^2 + \cdots + E_n x^n + \cdots$$

In this notation, the proposed problem requires to find E_{100}. Instead of our single original unknown E_{100}, we have introduced infinitely many new unknowns: we should find A_n, B_n, C_n, D_n, and E_n for $n = 0, 1, 2, 3, \ldots$. Yet, for some of these, the result is well known or obvious.

$$A_0 = A_1 = A_2 = \cdots = A_n = \cdots = 1$$
$$B_0 = C_0 = D_0 = E_0 = 1$$

Moreover, the unknowns introduced are not unrelated:

$$A_0 + A_1 x + A_2 x^2 + \cdots = (B_0 + B_1 x + B_2 x^2 + \cdots)(1 - x^5)$$

from which we conclude, by looking at the coefficient of x^n, that

$$A_n = B_n - B_{n-5}$$

Find analogous relations and find the intermediaries through which the unknown E_{100} is connected with the values already known. Eventually, you should obtain a numerical value for E_{100}.

†**3.85.** Find the nth derivative $y^{(n)}$ of the function $y = x^{-1} \log x$.

We obtain by straightforward differentiation and algebraic transformation

$$y' = - x^{-2} \log x + x^{-2}$$
$$y'' = 2x^{-3} \log x - 3x^{-3}$$
$$y''' = -6x^{-4} \log x + 11x^{-4}$$

and from these (or more) cases we may guess that the desired nth derivative is of the form

$$y^{(n)} = (-1)^n n! x^{-n-1} \log x + (-1)^{n-1} c_n x^{-n-1}$$

where c_n is an integer depending on n (but independent of x). Prove this and express c_n in terms of n.

3.86. Find a short expression for
$$1 + 2x + 3x^2 + \cdots + nx^{n-1}$$
(Do you know a related problem? Could you use its result—or its method?)

3.87. Find a short expression for
$$1 + 4x + 9x^2 + \cdots + n^2x^{n-1}$$
(Do you know a related problem? Could you use its result—or its method?)

3.88. (Continued) Generalize.

3.89. Being given that
$$a_{n+1} = a_n \frac{n + \alpha}{n + 1 + \beta}$$
for $n = 1, 2, 3, \ldots$ and $\alpha \neq \beta$, show that
$$a_1 + a_2 + a_3 + \cdots + a_n = \frac{a_n(n + \alpha) - a_1(1 + \beta)}{\alpha - \beta}$$

3.90. Find
$$\frac{p}{q} + \frac{p}{q}\frac{p+1}{q+1} + \frac{p}{q}\frac{p+1}{q+1}\frac{p+2}{q+2} + \cdots + \frac{p}{q}\frac{p+1}{q+1}\frac{p+2}{q+2}\cdots\frac{p+n-1}{q+n-1}$$

3.91. *On the number π.* We consider the unit circle (its radius $= 1$); we circumscribe about it, and inscribe in it, a regular polygon with n sides; let C_n (circumscribed) and I_n (inscribed) stand for the perimeters of these two polygons, respectively.

Introduce the abbreviations
$$\frac{a+b}{2} = A(a, b), \qquad \sqrt{ab} = G(a, b), \qquad \frac{2ab}{a+b} = H(a, b)$$
(arithmetic, geometric, and harmonic mean, respectively).

(1) Find C_4, I_4, C_6, I_6.
(2) Show that
$$C_{2n} = H(C_n, I_n), \qquad I_{2n} = G(I_n, C_{2n})$$
(Thus, starting from C_6, I_6 we can compute the sequence of numbers
$$C_6, I_6; \qquad C_{12}, I_{12}; \qquad C_{24}, I_{24}; \qquad C_{48}, I_{48}; \qquad \ldots$$
by recursion as far as we wish, and so we can enclose π between two numerical bounds whose difference is arbitrarily small. Archimedes, in computing the first ten numbers of the sequence, that is, proceeding to regular polygons with 96 sides, found that
$$3\tfrac{10}{71} < \pi < 3\tfrac{1}{7}$$
See his *Works*, edited by T. L. Heath (Dover), pp. 91–98.)

3.92. *More problems.* Devise some problems similar to, but different from, the problems proposed in this chapter—especially such problems as you can solve.

CHAPTER 4

SUPERPOSITION

4.1. Interpolation

We need several steps to arrive at the final formulation of our next problem.

(1) We are given n different abscissas

$$x_1, \qquad x_2, \qquad x_3, \qquad \cdots, \qquad x_n$$

and n corresponding ordinates

$$y_1, \qquad y_2, \qquad y_3, \qquad \ldots, \qquad y_n$$

and so we are given n different points

$$(x_1, y_1), \qquad (x_2, y_2), \qquad (x_3, y_3), \qquad \ldots, \qquad (x_n, y_n)$$

We are required to find a function $f(x)$ the values of which at the given abscissas are the corresponding ordinates:

$$f(x_1) = y_1, \quad f(x_2) = y_2, \quad f(x_3) = y_3, \quad \ldots, \quad f(x_n) = y_n$$

In other words, we are required to find a curve, with the equation $y = f(x)$, that passes through the n given points; see Fig. 4.1. This is the problem of *interpolation*. Let us explore the background of this problem; such exploration may increase our interest in it and so our chances to solve it.

(2) The problem of interpolation may arise whenever we consider a quantity y depending on another quantity x. Let us take a more concrete case: let x be the temperature and y the length of a homogeneous rod, kept under constant pressure. To each temperature x there corresponds a certain length y of the rod; this is what we express by saying that y *depends*

99

100   PATTERNS

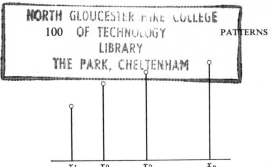

Fig. 4.1. Interpolation.

on x, or y is a *function* of x, or by writing $y = f(x)$. A physicist, in investigating experimentally the dependence of y on x, subjects the rod to different temperatures

$$x_1, \quad x_2, \quad x_3, \quad \ldots, \quad x_n$$

and, by measuring the length of the rod at each of these temperatures, he finds the values

$$y_1, \quad y_2, \quad y_3, \quad \ldots, \quad y_n$$

respectively. The physicist, of course, would like to know the length y also at some such temperature x as he has not yet had the opportunity to observe. That is, the physicist wants to know, on the basis of his n observations, the function $y = f(x)$ in its full extent, for the whole range of the independent variable x—and so he poses the problem of interpolation.

(3) Let us remark parenthetically that the physicist's problem is, in fact, more complicated. His values $x_1, y_1, x_2, y_2, \ldots, x_n, y_n$ are not the "true values" of the quantities measured but are affected by unavoidable errors of measurement. Therefore, his curve need not pass *through*, it should only pass *close* to, the given points.

Moreover, it is usual to distinguish two cases: the hitherto unobserved abscissa x, to which the physicist wants to find the corresponding ordinate y, may lie in the interval *between* the extreme observed values (x_1 and x_n in Fig. 4.1) or it may lie *outside* this interval: in the first case it is customary to speak of *interpolation* and in the latter of *extrapolation*. (It is usual to regard interpolation as more reliable than extrapolation.)

Yet let us disregard this distinction and the other remarks of this subsection, for the time being; let us close the parenthesis and return to the standpoint of the subsections (1) and (2).

(4) The problem posed in subsection (1) is utterly indeterminate: there is an inexhaustible variety of curves passing through the n given points. His n observations, by themselves, do not entitle the physicist to prefer one of those curves to the others. If the physicist decides to draw a curve, he

must have some reason for his choice *outside* his n observations—what reason?

Thus the problem of interpolation raises (and this adds a good deal to its interest) a general question: What suggests, or what justifies, the transition to a mathematical formulation from given observations and a given mental background? This is a major philosophical question—yet, as it is rather unlikely that major philosophical questions can be satisfactorily answered, we turn to another aspect of the problem of interpolation.

(5) It would be natural to modify the problem stated in subsection (1) by asking for the *simplest* curve passing through the n given points. This modification, however, leaves the problem indeterminate, even vague, since "simplicity" is hardly an objective quality: we may judge simplicity according to our personal taste, standpoint, background, or mental habits.

Yet, in the case of our problem, we may give an interpretation to the term "simple" that looks acceptable and leads to a determinate and useful formulation. First, let us regard addition, subtraction, and multiplication as the simplest computational operations. Then, let us regard such functions as the simplest the values of which can be computed by the simplest computational operations. Accepting both points, we have to regard the polynomials as the simplest functions; a polynomial is of the form

$$a_0 + a_1 x + a_2 x^2 + \cdots + a_n x^n$$

Its value can be computed by the three simplest computational operations from the numerically given coefficients $a_0, a_1, \ldots, a_n$ and the value of the independent variable x. If we assume that $a_n \neq 0$, the degree of the polynomial is n.

Finally, being given two polynomials of different degree, let us regard the one with the lower degree as the simpler. If we accept this point too, the problem of passing the simplest possible curve through n points becomes a determinate problem, the problem of *polynomial interpolation*, which we formulate as follows:

Being given n different numbers $x_1, x_2, \ldots, x_n$ and n further numbers $y_1, y_2, \ldots, y_n$, find the polynomial $f(x)$ of the lowest possible degree satisfying the n conditions

$$f(x_1) = y_1, \quad f(x_2) = y_2, \quad \ldots, \quad f(x_n) = y_n$$

4.2. A special situation

If we see no other approach to the proposed problem, we may try to *vary the data*. For instance, we may keep one given ordinate fixed and let the others decrease; and so we may hit upon a special situation that looks more accessible.   We need not touch the given abscissas, we accept any n different numbers

$$x_1, \quad x_2, \quad x_3, \quad \ldots, \quad x_n$$

but we choose a particularly simple system of ordinates:

$$0, \quad 1, \quad 0, \quad \ldots, \quad 0$$

respectively.   (All given ordinates vanish, except the one corresponding to the abscissa x_2; see Fig. 4.2.)

There is an interesting piece of information: the polynomial assuming these values vanishes at $n - 1$ given points, has the $n - 1$ different roots $x_1, x_3, x_4, \ldots, x_n$, and, therefore, it must be divisible by each of the following $n - 1$ factors:

$$x - x_1, \quad x - x_3, \quad x - x_4, \quad \ldots, \quad x - x_n$$

Therefore, it must be divisible by the product of these $n - 1$ factors, and so it is at least of degree $n - 1$.   If the polynomial attains this lowest possible degree $n - 1$, it must be of the form

$$f(x) = C(x - x_1)(x - x_3)(x - x_4)\ldots(x - x_n)$$

where C is some constant.

Have we used all the data?   There remains the ordinate corresponding to the abscissa x_2 to be taken into account:

$$f(x_2) = C(x_2 - x_1)(x_2 - x_3)(x_2 - x_4)\ldots(x_2 - x_n) = 1$$

We compute C from this equation, substitute the value computed for C in the expression of $f(x)$, and find so

$$f(x) = \frac{(x - x_1)\,(x - x_3)\,(x - x_4)\ldots(x - x_n)}{(x_2 - x_1)(x_2 - x_3)(x_2 - x_4)\ldots(x_2 - x_n)}$$

Obviously, this polynomial $f(x)$ takes the required values for all given abscissas.   We have succeeded in solving the problem of polynomial interpolation in a particular case, in a special situation.

4.3. Combining particular cases to solve the general case

We were lucky to spot such an especially accessible particular case.   To

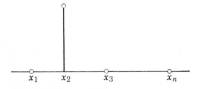

Fig. 4.2. A special situation.

deserve our luck, we should try now to make good use of the solution obtained.

By modifying a little the solution obtained, we can handle a slightly more extended particular case: to the given abscissas

$$x_1, \quad x_2, \quad x_3, \quad \ldots, \quad x_n$$

we let correspond the ordinates

$$0, \quad y_2, \quad 0, \quad \ldots, \quad 0$$

respectively. We obtain the polynomial that assumes these values by multiplying the expression obtained in sect. 4.2 by an obvious factor:

$$y_2 \frac{(x - x_1)(x - x_3)(x - x_4)\ldots(x - x_n)}{(x_2 - x_1)(x_2 - x_3)(x_2 - x_4)\ldots(x_2 - x_n)}$$

In this expression, the abscissa x_2 plays a special role, distinct from the common role that falls to the other abscissas. Yet there is no peculiar virtue in the abscissa x_2: we can let any other given abscissa play that special role. And so, if to the abscissas

$$x_1, \quad x_2, \quad x_3, \quad \ldots, \quad x_n$$

we let correspond the values displayed on any one of the n following lines

$$
\begin{array}{ccccc}
y_1, & 0, & 0, & \ldots, & 0 \\
0, & y_2, & 0, & \ldots, & 0 \\
0, & 0, & y_3, & \ldots, & 0 \\
. & . & . & . & . \\
0, & 0, & 0, & \ldots, & y_n
\end{array}
$$

we can write down an expression for the polynomial of degree $n - 1$ that assumes the values on that line at the corresponding abscissas.

We have here outlined the solution in n different particular cases of our problem. Can you combine them so as to obtain the solution of the general case from the combination? Of course, you can, by adding the n expressions outlined:

$$
\begin{aligned}
f(x) = {} & y_1 \frac{(x - x_2)(x - x_3)(x - x_4)\ldots(x - x_n)}{(x_1 - x_2)(x_1 - x_3)(x_1 - x_4)\ldots(x_1 - x_n)} \\
& + y_2 \frac{(x - x_1)(x - x_3)(x - x_4)\ldots(x - x_n)}{(x_2 - x_1)(x_2 - x_3)(x_2 - x_4)\ldots(x_2 - x_n)} \\
& + y_3 \frac{(x - x_1)(x - x_2)(x - x_4)\ldots(x - x_n)}{(x_3 - x_1)(x_3 - x_2)(x_3 - x_4)\ldots(x_3 - x_n)} \\
& \quad . \quad . \quad . \quad . \quad . \quad . \quad . \quad . \quad . \quad . \quad . \\
& + y_n \frac{(x - x_1)(x - x_2)(x - x_3)\ldots(x - x_{n-1})}{(x_n - x_1)(x_n - x_2)(x_n - x_3)\ldots(x_n - x_{n-1})}
\end{aligned}
$$

is a polynomial, of degree not exceeding $n - 1$, that satisfies the condition

$$f(x_i) = y_i \qquad \text{for} \quad i = 1, 2, 3, \ldots, n$$

as we can see at a glance if we realize the structure of its expression. (Are there any questions?)

4.4. The pattern

The foregoing solution of the interpolation problem, which is due to Lagrange, has a highly suggestive general plan. *Have you seen it before?*

(1) The reader is probably familiar with, and by the foregoing may be reminded of, the usual proof of a well-known theorem of plane geometry: "The angle at the center of a circle is double the angle at the circumference on the same base, that is, on the same arc." (The arc is emphasized by a double line in Figs. 4.3 and 4.4.) The proof is based on two remarks, and proceeds in two steps; cf. Euclid III 20.

(2) There is a more accessible *special situation*: If one of the sides of the angle at the circumference is a diameter, see Fig. 4.3, the angle at the center α is obviously the sum of two angles of an isosceles triangle; these two angles are equal to each other, and one of them is the angle at the circumference, β. This proves the desired equation

$$\alpha = 2\beta$$

for the special situation of Fig. 4.3.

(3) Now, we have no more the special situation of Fig. 4.3 before us. We can, however, draw a diameter (dotted line in Fig. 4.4) through the vertex of the angle at the circumference, and then the special situation arises twice in the figure. Let the equations

$$\alpha' = 2\beta', \qquad \alpha'' = 2\beta''$$

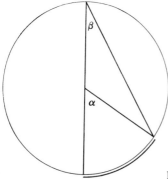

Fig. 4.3. A special situation.

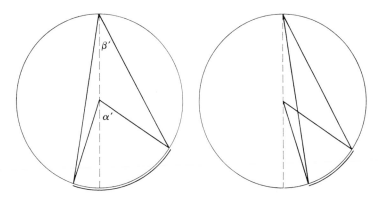

Fig. 4.4. The general case.

refer to these special situations (see Fig. 4.3). These equations are firmly established by the considerations of subsection (2). The angles α and β, at the center and at the circumference, respectively, with which the desired theorem deals, can be exhibited as sum or as difference, according as we have one or the other case represented by Fig. 4.4 before us:

$$\alpha = \alpha' + \alpha'', \quad \beta = \beta' + \beta'' \qquad \text{or} \qquad \alpha = \alpha' - \alpha'', \quad \beta = \beta' - \beta''$$

Now, by adding and subtracting our two already established equations, we obtain

$$\alpha' + \alpha'' = 2(\beta' + \beta''), \qquad \alpha' - \alpha'' = 2(\beta' - \beta'')$$

respectively, and this proves the desired theorem

$$\alpha = 2\beta$$

in full generality.

(4) Now, let us compare the two problems discussed in this chapter: the problem to find, from algebra, treated in sects. 4.1, 4.2, and 4.3; and the problem to prove, from plane geometry, treated in subsections (1), (2), and (3) of the present section. Although these problems differ in several respects, their solutions show the same *pattern*. In both examples, the result was attained in two steps.

First, we were lucky enough to spot a particularly accessible case, a *special situation*, and gave a solution well adapted, but restricted, to this special situation; see sect. 4.2 and subsection (2), Fig. 4.2 and Fig. 4.3.

Then, by *combining particular cases* to which the restricted solution is applicable, we obtained the full, unrestricted solution, applicable to the *general case*; see sect. 4.3 and subsection (3).

Let us introduce two terms which underline certain features of this pattern.

The first step deals with a particular case which is not only especially accessible, but also especially useful; we can appropriately call it a *leading particular case*: it leads the way to the general solution.[1]

The second step combines particular cases by a specific algebraic operation. In sect. 4.3, *n* particular solutions, after being multiplied by given constants, are added to form the general solution. In subsection (3), we add and subtract equations dealing with the special situation to obtain the general proof. Let us call the algebraic operation employed in sect. 4.3 [there is more generality there than in subsection (3)] *linear combination* or *superposition*. (More about this concept in ex. 4.11.)

We may use the terms introduced to outline our pattern: *Starting from a leading special situation we attain the general solution by superposition of particular cases.*

Other comments and more examples may enable the reader to fill in this outline. He may even burst this outline and enlarge the scope of the pattern.

Examples and Comments on Chapter 4

First Part

4.1. In proving the expression $Bh/3$ for the volume of a pyramid (B is the base, h the height) we may regard the case of the tetrahedron (which is a pyramid with triangular base) as leading particular case and use superposition. How?

4.2. If $f(x)$ is a polynomial of degree k, there exists a polynomial $F(x)$ of degree $k + 1$ such that, for $n = 1, 2, 3, \ldots,$

$$f(1) + f(2) + f(3) + \cdots + f(n) = F(n)$$

In proving this theorem, we may regard the result of ex. 3.3 as leading particular case and use superposition. How?

4.3. (Continued) There is, however, another way: we may regard the result of ex. 3.34 as leading particular case and give so, by using superposition, a different proof. How?

4.4. Being given the coefficients $a_0, a_1, a_2, \ldots, a_k$, find numbers $b_0, b_1, b_2, \ldots, b_k$ such that the equation

$$a_0 x^k + a_1 x^{k-1} + \cdots + a_k = b_0 \binom{x}{k} + b_1 \binom{x}{k-1} + \cdots + b_k \binom{x}{0}$$

holds identically in x (notation of ex. 3.65).

Show that this problem has just one solution.

[1] MPR, vol. 1, p. 24, ex. 7, 8, 9.

4.5. By applying the method of ex. 4.3, give a new derivation for the expression of S_3 obtained in sect. 3.3.

4.6. By applying the result of ex. 4.3 (the theorem stated in ex. 4.2) give a new derivation for the expression of S_3 obtained in sect. 3.3.

4.7. What does ex. 4.3 yield for the problem of ex. 3.3?

4.8. A question on sect. 4.1: What about the particular case $n = 2$? When just two points are given, it would be natural to say that the simplest line passing through them is the straight line, which is uniquely determined. Is this in agreement with the standpoint at which we eventually arrived in sect. 4.1(5)?

4.9. A question on sect. 4.2: What about the particular case in which

$$y_i = 0 \quad \text{for} \quad i = 1, 2, \ldots n$$

that is, all the given ordinates vanish?

4.10. A question on sect. 4.3: Does the polynomial $f(x)$ obtained satisfy all clauses of the condition? Is its degree the lowest possible?

4.11. *Linear combination or superposition.* Let

$$V_1, \quad V_2, \quad V_3, \quad \cdots, \quad V_n$$

be n mathematical objects of some clearly stated nature (belonging to some clearly defined set) such that their *linear combination*

$$c_1 V_1 + c_2 V_2 + c_3 V_3 + \cdots + c_n V_n$$

formed with any n numbers

$$c_1, \quad c_2, \quad c_3, \quad \cdots, \quad c_n$$

is of the same nature (belongs to the same set).

Here are two examples.

(a) V_1, V_2, $V_3, \ldots, V_n$ are polynomials of degree not exceeding a certain given number d; their linear combination is again a polynomial of degree not exceeding d.

(b) V_1, V_2, $V_3, \ldots, V_n$ are vectors parallel to a given plane; their linear combination (addition means here vector-addition) is again a vector parallel to the given plane.

Example (a) plays a role in sect. 4.3. With regard to sect. 4.4(3) let us observe that addition and subtraction are special cases of linear combination ($n = 2$; $c_1 = c_2 = 1$ and $c_1 = -c_2 = 1$, respectively).

Example (b) is suggestive; such objects as can be linearly combined subject to the "usual" laws of algebra are called "vectors," and their set is called a "vector-space," in abstract algebra.

Linear combinations (vector-spaces) play a role in several advanced branches of mathematics. We can consider here only a few not too advanced examples (ex. 4.12, 4.13, 4.14, and 4.15).

We use here the two terms "linear combination" and "superposition" in the same meaning, but we use the latter term more often. The term "superposition" is often employed in physics (especially in wave theory). We take here just one example from physics, ex. 4.16, which is simple enough for us and important in several respects.

†4.12. *Homogeneous linear differential equations with constant coefficients.* Such an equation is of the form

$$y^{(n)} + a_1 y^{(n-1)} + \cdots + a_{n-1} y' + a_n y = 0$$

$a_1, a_2, \ldots, a_n$ are given numbers, called the *coefficients* of the equation; n is the *order* of the equation; y is a function of the independent variable x; $y', y'', \ldots, y^{(n)}$ denote, as usual, the successive derivatives of y. A function y satisfying the equation is called a *solution*, or an "integral."

(*a*) Show that a linear combination of solutions is a solution.

(*b*) Show that there is a particular solution of the special form

$$y = e^{rx}$$

where r is an appropriately chosen number.

(*c*) Combine particular solutions of such special form to obtain a solution as general as possible.

†4.13. Find a function y satisfying the differential equation

$$y'' = -y$$

and the initial conditions

$$y = 1, \quad y' = 0 \quad \text{for} \quad x = 0$$

4.14. *Homogeneous linear difference equations with constant coefficients.* Such an equation is of the form

$$y_{k+n} + a_1 y_{k+n-1} + \cdots + a_{n-1} y_{k+1} + a_n y_k = 0$$

$a_1, a_2, \ldots, a_n$ are given numbers, called the *coefficients* of the equation; n is the *order* of the equation; an infinite sequence of numbers

$$y_0, y_1, y_2, \ldots, y_k, \ldots$$

which satisfies the equation for $k = 0, 1, 2, \ldots$ is called a *solution*.

(We may regard y_x as a function of the independent variable x defined for non-negative integral values of x. On the other hand, we may regard the proposed equation as a recursion formula, that is, a uniform rule by virtue of which we can compute any term y_{k+n} of the sequence from n foregoing terms $y_{k+n-1}, y_{k+n-2}, \ldots, y_k$—or y_k from $y_{k-1}, y_{k-2}, \ldots, y_{k-n}$.)

(*a*) Show that a linear combination of solutions is a solution.

(*b*) Show that there is a particular solution of the special form

$$y_k = r^k$$

where r is an appropriately chosen number.

(c) Combine particular solutions of such special form to obtain a solution as general as possible.

4.15. The sequence of *Fibonacci numbers*

$$0, 1, 1, 2, 3, 5, 8, 13, \ldots$$

is defined by the difference equation (recursion formula)

$$y_k = y_{k-1} + y_{k-2}$$

valid for $k = 2, 3, 4, \ldots$ and the initial conditions

$$y_0 = 0, \quad y_1 = 1$$

Express y_k in terms of k.

4.16. *Superposition of motions.* Galileo, having found the law of falling bodies and the law of inertia, combined these laws to discover the trajectory of (the curve described by) a projectile. The reader who realizes how much he is helped by modern notation may relive, in a fashion, this discovery of Galileo.

Let x and y denote rectangular coordinates in a vertical plane; the x axis is horizontal, the y axis points upward. A projectile (a material point devoid of friction) moves in this plane starting from the origin at the instant $t = 0$; t is the time. The initial velocity of the projectile is v, its initial direction includes the angle α with the positive x axis. With the actual motion of the projectile, we may associate three virtual motions, starting from the same point at the same time.

(a) A heavy material point, starting from rest and falling freely, has, at the time t, the coordinates

$$x_1 = 0, \qquad y_1 = -\tfrac{1}{2}gt^2$$

(b) A material point free from gravity, which has received the vertical component $v \sin \alpha$ of the initial velocity, has, at the time t, by virtue of the law of inertia, the coordinates

$$x_2 = 0, \qquad y_2 = tv \sin \alpha$$

(c) A material point, free from gravity, which has received the horizontal component of the initial velocity, has, at the time t, by virtue of the law of inertia, the coordinates

$$x_3 = tv \cos \alpha, \qquad y_3 = 0$$

If the actual motion is compounded from these three virtual motions according to the "simplest" assumption, what is its trajectory?

Second Part

Opportunity is offered to the reader to participate in an investigation, important phases of which are indicated by ex. 4.17 and ex. 4.24.

4.17. *The multiplicity of approaches.* In a tetrahedron two opposite edges have the same length a, they are perpendicular to each other, and each is perpendicular to the line of length b that joins their midpoints. Find the volume of the tetrahedron.

There are several different approaches to this problem. The reader who needs help may look at some, or all, of the following ex. 4.18–23. If he wishes to visualize the spatial relations involved, he may look for a simple orthogonal projection, or for a simple cross-section.

4.18. *What is the unknown?* The unknown in ex. 4.17 is the volume of a tetrahedron.

How can you find this kind of unknown? The volume of a tetrahedron can be computed if its base and height are given—but neither of these two quantities is given in ex. 4.17.

Well, *what is the unknown?*

4.19. (Continued) You need the area of a triangle—*how can you find this kind of unknown?* The area of a triangle can be computed if its base and height are given—but only one of these two quantities is given for the triangle that forms the base of the tetrahedron of ex. 4.17.

You need the length of a line—*how can you find this kind of unknown?* The usual thing is to compute the length of a line from some triangle—but, in the figure, there is no triangle in which the height of the tetrahedron of ex. 4.17 would be contained.

In fact, there is no such triangle; but *could you introduce one?* At any rate, *introduce suitable notation* and collect whatever you have.

4.20. *Here is a problem related to yours and solved before*: "The volume of a tetrahedron can be computed if its base and height are given." You cannot apply this to ex. 4.17 immediately, because the base and height of that tetrahedron are not given. There may be, however, other, more accessible tetrahedra *around.*

4.21. (Continued) And there may be more accessible tetrahedra *within.*

4.22. *More knowledge* may help. Ex. 4.17 is easy for you, if you know the *prismoidal formula.*

A *prismoid* is a polyhedron. Two faces of the prismoid, called the *lower base* and the *upper base*, are parallel to each other; all the other faces are *lateral faces.* The prismoid has three kinds of edges: edges surrounding the lower base, edges surrounding the upper base, and *lateral edges.* Any lateral edge of the prismoid (this is an important part of its definition) joins a vertex of the lower base to a vertex of the upper base. A prism is a special prismoid.

The distance between the two bases is the *height* of the prismoid. A plane that is parallel to, and equidistant from, the two bases, the lower and the upper, intersects the prismoid in a polygon called the *midsection.*

Let V stand for the volume of the prismoid, h for the height,

$$L, \quad M, \quad \text{and} \quad N$$

for the areas of its lower base, its midsection, and its upper base, respectively. Then (this expression for V is called the prismoidal formula)

$$V = \frac{(L + 4M + N)h}{6}$$

Apply it to ex. 4.17.

4.23. Perhaps you have abandoned the path to the solution of ex. 4.17 that starts from ex. 4.18 and leads through ex. 4.19, but attained the result following some other route. If so, look at the result, return to the abandoned path, and follow it to the end.

4.24. *The prismoidal formula.* Study all sides of the question, consider it under various aspects, turn it over and over in your mind—we did so, look at Fig. 4.5. Having found four different derivations for the same result, we should be able to profit from their comparison.[2]

Out of our four derivations, three do not use the prismoidal formula, but one does (ex. 4.22). Hence, for that particular case of the prismoidal formula that intervenes in the problem treated, we have in fact, implicitly at least, three different proofs. Could we render one of these proofs explicit and extend it so that it proves that formula not only in a particular case, but generally?

On the face of it, which one of the three derivations in question (ex. 4.20, ex. 4.21, and ex. 4.18, 4.19, 4.23) appears to have the best chances?

4.25. Verify the prismoidal formula for the prism (which is a very special prismoid).

4.26. Verify the prismoidal formula for the pyramid (which, in an appropriate position, can be regarded as a prismoid—a degenerate, or limiting, case of a prismoid if you prefer, with an upper base shrunken into a point).

4.27. In generalizing the situation that underlies the solution of ex. 4.20, we consider a prismoid P split into n prismoids $P_1, P_2, \ldots, P_n$ which are nonoverlapping and fill P completely, their lower bases fill the lower base of P, their upper bases the upper base of P. (In the case of ex. 4.20, Fig. 4.5b, P is a prism with a square base, $n = 5$, P_1, P_2, P_3, and P_4 are congruent tetrahedra,

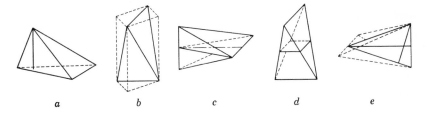

a b c d e

Fig. 4.5. Turn it over and over, consider it under various aspects, study all sides.

[2] In this, we follow Leibnitz's opinion; see the quotation preceding ex. 3.31.

P_5 another tetrahedron.)   Show: If the prismoidal formula is valid for n prismoids out of the $n + 1$ considered, it is necessarily valid also for the remaining one.

4.28. In generalizing the situation that underlies the solution of ex. 4.22, Fig. 4.5*d*, we let l and n denote opposite edges of a tetrahedron (l lower, n upper).   Pass a plane through l parallel to n, and another plane through n parallel to l; let h denote the distance of these two (parallel) planes.   The tetrahedron can be regarded as a prismoid (a degenerate prismoid, if you prefer) of which the edges l and n are the bases, the lower and the upper, respectively, and h the height.   (The midsection is a parallelogram.)

Verify the prismoidal formula for this kind of prismoid.

4.29. Prove the prismoidal formula generally (by superposition of particular cases treated previously).

4.30. *No chain is stronger than its weakest link.*   Reexamine the solution of ex. 4.28.

4.31. Reexamine the solution of ex. 4.29.

†**4.32.** *Simpson's rule.*   Let $f(x)$ be a function defined (and continuous) in the interval

$$a \leqq x \leqq a + h$$

put

$$\int_a^{a+h} f(x)dx = I$$

$$f(a) = L, \qquad f\left(a + \frac{h}{2}\right) = M, \qquad f(a + h) = N$$

Then, under certain conditions which we intend to explore,

$$I = \frac{L + 4M + N}{6} h$$

this expression for I is called *Simpson's rule.*

Let n denote a non-negative integer, take

$$f(x) = x^n, \qquad a = -1, \qquad h = 2$$

and determine those values of n for which the expression of the integral I by Simpson's rule is valid.

(Even when Simpson's rule is not exactly valid, it may be "approximately valid," that is, the difference between the two sides may be relatively small. This is frequently the case, and hence Simpson's rule is important for the approximate evaluation of integrals.)

†**4.33.** Prove that Simpson's rule is valid for any polynomial of degree not exceeding 3, provided that $a = -1$ and $h = 2$.

†**4.34.** Prove that Simpson's rule is valid for any polynomial of degree not exceeding 3 and unrestricted a and h.

†**4.35.** Derive the prismoidal formula from ex. 4.34, using solid analytic geometry and integral calculus. ("To appreciate the easy way do it first the hard way," said the traditional mathematics professor.)

4.36. *Widening the scope.* In solving some of the foregoing problems we actually went beyond the outline of the pattern of superposition formulated in sect. 4.4(4). We have, in fact, attained the general solution by superposing more accessible particular cases, but these particular cases were not all of the same type, they did not all belong to the same special situation. (In the solution of ex. 4.29, some of the superposed solids are pyramids, treated in ex. 4.26, others are tetrahedra in a special position, treated in ex. 4.28. Also in the solution of ex. 4.33, we superpose cases of different nature.) Essentially, we deviated from the formulation of sect. 4.4(4) in just one point: we did not start from one leading special situation, but from several such situations. Let us, therefore, enlarge the outline of our pattern: *Starting from a leading special situation, or from several such situations, we attain the general solution by superposition of particular cases.*

The pattern of superposition points out a path from a leading special case (or from a few such cases) to the general case. There is a very different connecting path between the same endpoints with which the ambitious problem-solver should be equally acquainted: it is often possible to reduce the general case to a leading special case by an appropriate *transformation.* (The general case of ex. 4.34 is reduced to the special case of ex. 4.33 by a transformation of the variable of integration.) For a suggestive discussion of this topic see J. Hadamard, *Leçons de géométrie élémentaire. Géométrie plane,* 1898; Méthodes de transformation, pp. 272–278.

PART TWO

TOWARD A GENERAL METHOD

Human wisdom remains always one and the same
although applied to the most diverse objects and it is
no more changed by their diversity than the sunshine
is changed by the variety of objects
which it illuminates.

DESCARTES: Rule I, *Œuvres*, vol. X, p. 360.

CHAPTER 5

PROBLEMS

*The solution of problems is the most characteristic and
peculiar sort of voluntary thinking.*
WILLIAM JAMES

5.1. What is a problem?

In what follows, the word "problem" will be taken in a very comprehensive meaning.   Our first task is to outline this meaning.

Getting food is usually no problem in modern life.   If I get hungry at home, I grab something in the refrigerator, and I go to a coffeeshop or some other shop if I am in town.   It is a different matter, however, when the refrigerator is empty or I happen to be in town without money; in such a case, getting food becomes a problem.   In general, a desire may or may not lead to a problem.   If the desire brings to my mind immediately, without any difficulty, some obvious action that is likely to attain the desired object, there is no problem.   If, however, no such action occurs to me, there is a problem.   Thus, to have a problem means: *to search consciously for some action appropriate to attain a clearly conceived, but not immediately attainable, aim.*   To solve a problem means to find such action.

A problem is a"great"problem ·if it is very difficult, it is just a"little" problem if it is just a little difficult.   Yet some degree of difficulty belongs to the very notion of a problem: where there is no difficulty, there is no problem.

A typical problem is to find the way to a preassigned spot in some little known region.   We can easily imagine how serious this problem was for our primitive ancestors who dwelt in a primeval forest.   This may or may not be the reason that the solution of any problem appears to us somehow as finding a way: a way out of a difficulty, a way around an obstacle.

The greater part of our conscious thinking is concerned with problems. When we do not indulge in mere musing or daydreaming, our thoughts are directed towards some end, we seek ways and means to that end, we try to think of some course following which we could attain our aim.

Solving problems is the specific achievement of intelligence, and intelligence is the specific gift of man. The ability to go round an obstacle, to undertake an indirect course where no direct course presents itself, raises the clever animal above the dull one, raises man far above the most clever animals, and men of talent above their fellow men.

Nothing is more interesting for us humans than human activity. The most characteristically human activity is solving problems, thinking for a purpose, devising means to some desired end. Our aim is to understand this activity—it seems to me that this aim deserves a good deal of interest.

In the foregoing, we have studied elementary mathematical problems, in grouping together problems accessible to the same method of solution. We have acquired so a certain experimental basis and, from this basis, we shall now try to ascend to higher generality, attempting to embrace, as far as possible, also nonmathematical problems. To aim at a general method applicable to all sorts of problems may seem too ambitious, but it is quite natural: although the variety of problems we may face is infinite, each of us has just one head to solve them, and so we naturally desire just one method to solve them.

5.2. Classification of problems

A student is taking a written examination in mathematics; he is just an average student, but he did some work to prepare himself for this examination. After having read a proposed problem, he may ask himself: "What kind of problem is this?" In fact, it could be to his advantage to ask this question: if he can classify his problem, recognize its type, place it in such and such a chapter of his textbook, he has made some progress: he may now recall the method he has learned for solving this type of problem.

The same holds, in a sense, for all levels of problem solving. The question "What kind of problem is this?" leads to the next question "What can be done about this kind of problem?" and these questions may be asked with profit even in quite advanced research.

And so it may be useful to classify problems, to distinguish problems of various types. A good classification should introduce such types that *the type of problem may suggest the type of solution.*

We shall not enter now upon a detailed, or attempt a perfect, classification. In interpreting with some latitude a tradition which goes back to Euclid and his commentators, we just wish to characterize two very general types of problems.

Euclid's Elements contain axioms, definitions, and "propositions."

His commentators and some of his translators distinguish two kinds of "propositions": the aim of the first kind (the Latin name is "problema") is to construct a figure, the aim of the second kind (the Latin name is "theorema") is to prove a theorem.   In extending this distinction, we shall consider two kinds of problems, problems "to find" and problems "to prove."   The aim of a *problem to find* is to find (construct, produce, obtain, identify,...) a certain object, the unknown of the problem.   The aim of a *problem to prove* is to decide whether a certain assertion is true or false, to prove it or disprove it.

For instance, when you ask "What did he say?" you pose a problem to find.   Yet, when you ask "Did he say that?" you pose a problem to prove.

For more details about these two kinds of problems see the next two sections.

5.3. Problems to find

The aim of a "problem to find" is to find a certain object, the *unknown* of the problem, satisfying the *condition* of the problem which relates the unknown to the *data* of the problem.   Let us consider two examples.

"Being given two line-segments a and b, and the angle γ, construct the parallelogram of which the given line-segments are adjacent sides including the angle γ."

"Being given two line-segments a and b, and the angle γ, construct the parallelogram of which the given line-segments are the diagonals including the angle γ."

In both problems, the data are the same: the line-segments a and b, and the angle γ.   In both problems, the unknown is a parallelogram and so our problems are not *a priori* distinguishable by the nature of the unknown. What makes a difference between our two problems is the condition, the required relation between the unknown and the data: of course, the relation of the parallelogram to its sides differs from its relation to its diagonals.

The *unknown* may be of every imaginable category.   In a problem of geometric construction the unknown is a figure, for instance a triangle. When we are solving an algebraic equation, our unknown is a number, a root of that equation.   When we ask "What did he say?" the unknown may be a word, or a sequence of words, a sentence, or a sequence of sentences, a speech.   A problem clearly stated must specify the category (the set) to which the unknown belongs; we have to know from the start what kind of unknown we are supposed to find: a triangle, or a number, or a word, or....

A problem clearly stated must specify the *condition* that the unknown has to satisfy.   In the set of objects specified by the problem to which the unknown must belong, there is the subset of those objects that satisfy the condition, and any object belonging to this subset is called a *solution*. This subset may contain just one object: then the solution is unique. This subset may be empty: then there is no solution. (For remarks on the term "solution" see ex. 5.13.) We observe here that a problem to find can be taken in various meanings. Taken in a strict sense, the problem demands to find (produce, construct, identify, list, characterize,...) *all* the solutions (the entire subset mentioned above). Taken in a less strict sense, the problem may ask for just one (any one) solution, or some solutions. Sometimes it is enough to decide the existence of a solution, that is, to decide whether the set of solutions is empty or not. It is usual to take mathematical problems in the strict sense unless the contrary is explicitly stated, but for many practical problems the "strict sense" would make little sense.

When we deal with mathematical problems (unless the context hints the contrary) we shall use the phrase "the data" to denote all the given (known, granted,...) objects (or their full set) connected with the unknown by the condition. If the problem is to construct a triangle from its sides a, b, and c, the data are the three line-segments a, b, and c. If the problem is to solve the quadratic equation

$$x^2 + ax + b = 0$$

the data are the two given numbers a and b. A problem may have just one datum, or no data at all. Here is an example: "Find the ratio of the area of a circle to the area of a circumscribed square." The required ratio is independent of the size of the figure and so it is unnecessary to give the length of the radius or some other datum of this kind.

We shall call the unknown, the condition, and the data the *principal parts* of a problem to find. In fact, we can not reasonably hope to solve a problem that we do not understand. Yet, to understand a problem, we should know, and know very well, what is the unknown, what are the data, and what is the condition. And so it is advisable when we are working at a problem to pay especial attention to its principal parts.

5.4. Problems to prove

There is a rumor that, on a certain occasion, Secretary Soandso used an extremely crude expression (which we shall not reproduce here) in referring to Congressman Un Tel. It is just a rumor to which considerable doubt is attached. Yet the question "Did he say that?" has agitated many persons, was debated in the press, mentioned in a congressional committee, and may reach the courts. Whoever takes the question seriously has a "problem to prove" on his hands: he should lift the doubt about the rumor, he should prove the use of the alleged expression or disprove it, and proof or disproof should be supported by the best evidence available.

When we have a mathematical "problem to prove," we should lift the doubt about a clearly stated mathematical assertion A, we should prove A or disprove A. A celebrated unsolved problem of this kind is to prove or disprove Goldbach's conjecture: If the integer n is even and $n > 4$, then n is the sum of two odd primes.[1]

Goldbach's assertion (it is a mere assertion, we do not know yet whether it is true or false) is stated here in the *most usual form* of mathematical propositions: it consists of *hypothesis* and *conclusion*; the first part starting with "If" is the hypothesis, the second part starting with "then" is the conclusion.[2]

When we have to prove or disprove a mathematical proposition stated in the most usual form, the hypothesis and the conclusion of the proposition are appropriately called the *principal parts* of our problem. In fact, these principal parts deserve our especial attention. To prove the proposition we should discover a binding logical link between the principal parts, the hypothesis and the conclusion; to disprove the proposition we should show (by a counter-example, if possible) that one of the principal parts, the hypothesis, does not imply the other, the conclusion. Many mathematicians, great and small, tried to lift the doubt about Goldbach's conjecture, but without success: although very little knowledge is needed to understand the hypothesis and the conclusion, nobody has succeeded yet in linking them with a strict argument, and nobody has been able to produce a counter-example.

5.5. The components of the unknown, the clauses of the condition

If our problem is to construct a circle, we have to find, in fact, two things: the center of the circle and its radius. It may be advantageous to subdivide our task: of the two things wanted, the center and the radius, we may try to find first one then the other.

If our problem is to find a point in space and we use analytic geometry, we have to find, in fact, three numbers: the three coordinates x, y, and z of the point.

---

[1] MPR, vol. 1, pp. 4–5.

[2] There are mathematical propositions which cannot be naturally split into hypothesis and conclusion; see HSI, p. 155, Problems to find, problems to prove, 4. Here is a proposition of this kind: "In the decimal fraction of the number π there are nine consecutive digits 9." To prove or disprove this proposition is a definite mathematical problem—which seems to be, at present, hopelessly difficult. "A fool can ask more questions than nine wise men can answer."

According to the viewpoint which we prefer, we can say that, in our first example, there are two unknowns or just one unknown and, in our second example, there are three unknowns or just one unknown. There is, however, still another viewpoint which is often advantageous: we may say that, in both examples, there is just one unknown, but it is, in a sense, "subdivided." Thus, in our first example, the circle is the unknown, but it is a *bipartite* or *two-component* unknown; its *components* are its center and its radius. Similarly, in our second example, the point is a *tripartite* or *three-component* unknown; its components are its three coordinates x, y, and z. Generally, we may consider a *multipartite* or *multicomponent* unknown x having n components x_1, x_2, ..., x_n.

One advantage of the terminology we have just introduced is that, in certain general discussions, we need not distinguish between problems with one unknown and problems with several unknowns: in fact, we can reduce the latter case to the former by considering those several unknowns as components of one unknown. For instance, what we have said in sect. 5.3 applies essentially also to problems where we have to find several unknowns, although this case has not been explicitly mentioned in sect. 5.3. We shall see later that our terminology is useful in various contexts.

If our problem is a problem to find, there may be advantage in subdividing the condition into several parts or *clauses* as we have had ample opportunity to observe. In solving a problem of geometric construction, we may split the condition into two parts so that each part yields a locus for the unknown point (chapter 1). In solving a "word problem" by algebra, we split the condition into as many parts as there are unknowns, so that each part yields an equation (chapter 2).

If our problem is a problem to prove, there may be advantage in subdividing the hypothesis, or the conclusion, or both, into appropriate parts or clauses.

5.6. Wanted: a procedure

In constructing a figure in the style of Euclid's Elements, we are not free to choose our tools or instruments: we are supposed to construct the figure with ruler and compasses. Thus, the solution of the problem consists, in fact, in a sequence of *well-coordinated geometric operations* which start from the data and end in the required figure: our operations are drawing straight lines and circles, and determining their points of intersection.

This example may open our eyes and, looking sharper, we may perceive that the solution of many problems consists essentially in a *procedure*, a course of action, a scheme of well-interrelated operations, a *modus operandi*.

Take the problem of solving an equation of the second (or third, or fourth) degree. The solution consists in a scheme of well-coordinated *algebraic* operations which start from the data, the given coefficients of the equation, and end in the required roots: our operations are adding, subtracting, multiplying, or dividing given, or previously obtained, quantities, or extracting roots from such quantities.

Or consider a problem "to prove." The solution of the problem, the result of our efforts, is a proof, that is, a sequence of well-coordinated *logical* operations, of steps which start from the hypothesis and end in the desired conclusion of the theorem: each step infers some new point from appropriately chosen parts of the hypothesis, from known facts, or from points previously inferred.

Nonmathematical problems present a similar aspect. The builder of a bridge has to organize, coordinate, bring into a coherent scheme a tremendous multiplicity of operations: constructing approaches, shipping supplies, erecting scaffoldings, pouring concrete, riveting metallic parts, etc. etc. Moreover, he may be obliged to interrelate these operations with others of a very different nature: with financial, legal, even political transactions. All these operations depend on each other: most of them suppose that certain others have been previously performed.

Or take the case of the mystery story. The unknown is the murderer; the author tries to impress us by the performance of the detective-hero who devises a scheme, a course of action which, starting from the first indications, ends in recognizing and trapping the murderer.

The object of our quest may be an unknown of any nature or the discovery of the truth about any kind of question; our problem may be theoretical or practical, serious or trifling. To solve our problem, we have to devise a well-conceived, coherent scheme of operations, of logical, mathematical, or material operations proceeding from the hypothesis to the conclusion, from the data to the unknown, from the things we have to the things we want.

Examples and Comments on Chapter 5

5.1. Of a right prism with square base, find the volume V, being given the length a of a side of the base and the height h of the prism.

What is the unknown? What are the data? What is the condition?

5.2. Find two real numbers x and y satisfying the equation

$$x^2 + y^2 = 1$$

What is the unknown? What are the data? What is the condition? Describe the set of solutions.

5.3. Find two real numbers x and y satisfying the equation

$$x^2 + y^2 = -1$$

Describe the set of solutions.

5.4. Find two integers x and y satisfying the equation

$$x^2 + y^2 = 13$$

Describe the set of solutions.

5.5. Find three real numbers x, y, and z such that

$$|x| + |y| + |z| < 1$$

(1) Describe the set of solutions.

(2) Modify the problem by substituting $\leq$ for $<$, and describe the set of solutions for the modified problem.

5.6. State the theorem of Pythagoras.
What is the hypothesis? What is the conclusion?

5.7. Let n denote a positive integer and $d(n)$ the number of divisors of n (we mean positive integral divisors including 1 and n). For example

6 has the divisors 1, 2, 3, 6; $d(6) = 4$
9 has the divisors 1, 3, 9   ; $d(9) = 3$

Consider the proposition:
   $d(n)$ is odd or even according as n is, or is not, a square.
What is the hypothesis? What is the conclusion?

5.8. *To prove or to find?* Are the two numbers $\sqrt{3} + \sqrt{11}$ and $\sqrt{5} + \sqrt{8}$ equal? If they are not, which one is greater?

Restated in a generalized form, the problem is concerned with two numbers a and b, well defined by arithmetic operations, and requires of us to decide which one of the three possible cases

$$a = b, \qquad a > b, \qquad a < b$$

is actually true.

We may perceive different aspects of this problem.

(1) First, we have to prove or disprove the proposition $a = b$. If it turned out that this proposition is false, we have to prove or disprove the proposition $a > b$. We may tackle these two tasks also in the reverse order, or perhaps simultaneously; at any rate, we have here two problems to prove, linked with each other.

(2) There is a notation extensively used in various branches of mathematics: sgn x (read "the sign of x" or "signum x") is defined as follows:

$$\operatorname{sgn} x = \begin{cases} 1 & \text{when } x > 0 \\ 0 & \text{when } x = 0 \\ -1 & \text{when } x < 0 \end{cases}$$

The problem stated requires of us to find the number sgn $(a - b)$: this is a problem to find.

There is no formal contradiction (there need not be one if our terminology is carefully devised): in (1), we have a problem A consisting of two linked, simultaneously stated, problems to prove; in (2), we have a problem B, which is a problem to find. We do not regard these two problems A and B, stated in different terms, as *identical*—but they are *equivalent*. (This usage of the term "equivalent" is explained in HSI, Auxiliary problem 6, pp. 53–54, and will be explained again in Chapter 9.)

Moreover, there is no material disadvantage. On the contrary, it may be a good thing to see two different aspects of the same difficulty: one aspect may appeal more to us than the other, it may show a more accessible side and so give us a chance to attack the difficulty from that more accessible side.

5.9. *More problems.* Take any problem (there are many in the foregoing chapters), determine whether it is a problem "to find" or one "to prove," and ask accordingly:

What is the unknown? What are the data? What is the condition?

What is the conclusion? What is the hypothesis?

The aim of these questions here is just to familiarize you with the principal parts of problems. Yet, experience may show you that these questions, if seriously asked and carefully answered, are a great help in problem solving: in focusing your attention upon the principal parts of the problem, they deepen your understanding of the problem and they may start you in the right direction.

5.10. *The procedure of solution may consist of an unlimited sequence of operations.* If we are required to solve the equation

$$x^2 = 2$$

our task can be interpreted in various ways. The interpretation may be: "Find the positive square root of 2 to five significant figures"; in this case we fully discharge our duty in producing the decimal fraction 1.4142. Yet, the interpretation may also be "Extract the square root of 2" without any additional qualification or alleviation, and then we cannot perform our task by producing four, or any other given number of figures after the decimal point: the answer must be a *procedure*, a *scheme* of arithmetical operations that can yield *any* required number of decimal figures.

Here is another example: "Find the ratio of the area of the circle to the area of the circumscribed square." The answer is $\pi/4$ if we take the value of π for granted. Leibnitz gave the answer (expressed the ratio $\pi/4$) in form of an infinite series

$$\frac{1}{1} - \frac{1}{3} + \frac{1}{5} - \frac{1}{7} + \frac{1}{9} - \frac{1}{11} + \cdots$$

This series prescribes, in fact, a never ending sequence of arithmetical operations which can lead us to any number of figures in the decimal fraction of π

(in theory—in practice the procedure is much too slow). "Although this series, as it stands, is not suitable for a rapid approximation, but to present to the mind the ratio of the circle to the circumscribed square, I do not think that anything more suitable or more simple can be imagined," says Leibnitz.[3]

5.11. *Squaring the circle.* In solving a problem "to find" we seek an object, the "unknown" object, and very often we are led to seeking a procedure (a sequence of operations) to obtain that object—let us call the procedure sought, just for the sake of neater distinction, the "operational unknown." That a neat distinction is very desirable here may be illustrated by a historic example.

Being given the radius of a circle, construct *with ruler and compasses* a square having exactly the same area as the circle.

This is the strict form of the celebrated ancient problem of "squaring the circle" which originated with the early Greek geometers. We emphasize that the problem *prescribes the nature of the procedure* (of the "operational unknown"): we should construct a side of the desired square with straightedge and compasses, by drawing straight lines and circles, in using only points given or obtained by the intersection of previously drawn lines. And, of course, starting from the two endpoints of the given radius, we should attain the two endpoints of a side of the desired square in a *finite number* of steps.

After many centuries, in which an uncounted number of persons attempted the solution, it was proved (by F. Lindemann in 1882) that there is no solution: The square having the same area as the given circle undoubtedly "exists" (its side can be approximated to any given precision by various infinite processes known today, one of which is provided by the celebrated series of Leibnitz mentioned in ex. 5.10). Yet, a procedure of the desired kind (consisting of a finite sequence of operations with straightedge and compasses) does not exist. I wonder whether a clear distinction between the desired figure and the desired procedure, between the "unknown object" and the "operational unknown," would have diminished the number of unfortunate circle-squarers.

5.12. *Sequence and consequence.* In the construction of a bridge, fixing a prefabricated metallic piece in its proper place is an important operation. It may be essential that of two such operations one should precede the other (when the second piece cannot be fastened unless the first has been fastened before), but again it may be unessential (when the two pieces are independent). Thus, it may or may not be necessary to observe a definite sequence in the performance of two operations. Similarly, the steps of a proof are presented successively in a lecture or in print. Yet, a step may precede another step in time without preceding it in logic. We must distinguish between sequence and consequence, between succession in time and logical concatenation. (We shall return to this important matter in chapter 7.)

5.13. *An unfortunate ambiguity.* The word "solution" has several different meanings, some of which are very important and would deserve to be designated by an unambiguous term. In default of better ones, I propose a few such terms (adding a German equivalent to each).

---

[3] *Philosophische Schriften*, edited by Gerhardt, vol. IV, p. 278.

Solving object (Lösungsgegenstand) is an object satisfying the condition of the problem. If the aim of the problem is to solve an algebraic equation, a number satisfying the equation, that is, a root of the equation, is a solving object. Only a problem "to find" can have a solving object. A category (set) to which the solving object belongs must be specified in advance in a clearly stated problem—we must know in advance whether we seek a triangle, or a number, or what not. In fact, such specification (the precise designation of a set to which the unknown belongs) is a principal part of the problem. "To find the unknown" means to find (identify, construct, produce, obtain, . . .) the solving object (the set of all solving objects).

Solving procedure (Lösungsgang) is the procedure (the construction, the scheme of operations, the system of conclusions) that ends in finding the unknown of a problem to find, or in lifting the doubt about the assertion proposed by a problem to prove. Thus, "solving procedure" is a term applicable to both kinds of problems. At the beginning of our work, we do not know the solving procedure, the appropriate scheme of operations, but we are seeking it all the time in the hope that we shall know it at the end: this procedure is the aim of our quest, it is effectively, in a sense, our unknown, it is, let us say, our "operational unknown." (Cf. ex. 5.11.)

We could also talk about the "work of solving" (Lösungsarbeit) and the "result of solving" (Lösungsergebnis), but, in fact, I shall try not to appear too fussy and, except in a few important cases, I shall leave it to the reader to discover from the context what the word "solution" means in a given case: whether it means the object, the procedure, the result of the work, or the work itself.[4]

5.14. *Data and unknown, hypothesis and conclusion.* Euclid's Elements have a peculiar consistent style which some of us may be inclined to call solemn and others pedantic. All propositions are phrased according to a set pattern and, in this phrasing, the data and the unknown of a problem to find are so treated as if they were similar, or parallel, to the hypothesis and the conclusion of a problem, respectively. In fact, there is, as we shall see later, a certain similarity or parallelism between these principal parts of the two kinds of problems, which is of some importance from the problem-solver's viewpoint, and so from the viewpoint of our subject. Yet, it is inadmissible and illiterate to mix up the terms data and hypothesis or the terms unknown and conclusion and to apply any one of these terms to the kind of problem for which it is unfit. It is sad that such inadmissible and illiterate use of these important terms occurs sometimes even in print.

5.15. *Counting the data.* A triangle is determined by three sides, or by two sides and one angle (the included angle), or by one side and two angles, but it is not determined by three angles: 3 independent data are required to determine a triangle. (See also ex. 1.43 and 1.44.) To determine a polynomial of degree n in one variable (in x) $n + 1$ independent data are required: the $n + 1$ coefficients in the expansion of the polynomial in powers of x, or the $n + 1$

---

[4] Cf. HSI, p. 202; Terms, old and new, 8.

values taken by the polynomial at the points $x = 0, 1, 2, \ldots, n$, or at any other $n + 1$ different given points, and so on.   There are many important kinds of mathematical objects to determine which a definite number of independent data is required.   Therefore, when we are solving a problem to find, it is often advantageous to *count the data*, and to count them early.

5.16.   To determine a polygon with n sides

$$(n - 1) + (n - 2) = (n - 3) + n = 3 + 2(n - 3) = 2n - 3$$

independent data are required.   What do these four different expressions for the same number suggest to you?

5.17.   How many data are required to determine a pyramid the base of which is a polygon with n sides?

5.18.   How many data are required to determine a prism (which may be oblique) the base of which is a polygon with n sides?

5.19.   How many data are required to determine a polynomial of degree n in v variables?   (Its terms are of the form $cx_1^{m_1}x_2^{m_2}\ldots x_v^{m_v}$ where c is a constant coefficient and $m_1 + m_2 + \cdots + m_v \leqq n$.)

CHAPTER 6

WIDENING THE SCOPE

Divide each problem that you examine into as many parts as you can and as you need to solve them more easily.
DESCARTES: *Œuvres*, vol. VI, p. 18; Discours de la Méthode, Part II.

This rule of Descartes is of little use as long as the art of dividing...remains unexplained.... By dividing his problem into unsuitable parts, the unexperienced problem-solver may increase his difficulty.
LEIBNITZ: *Philosophische Schriften*, edited by Gerhardt, vol. IV, p. 331.

6.1. Wider scope of the Cartesian pattern

There are important methodical ideas involved in the Cartesian pattern that are not necessarily connected with the setting up of equations. The present chapter undertakes to disentangle some such ideas. We shall pass cautiously from equations to more general concepts. We begin with an example which is sufficiently general in some respects, but very concrete in another respect; it indicates the direction of our subsequent work.

(1) A certain problem has been translated into a system of four equations with four unknowns, x_1, x_2, x_3, and x_4. This system has a peculiar feature: not all of its equations involve all the unknowns. Now, we wish to emphasize just this feature: our notation will clearly show which equation involves which unknowns, but it will neglect other details. In fact, we write the four equations as follows:

$$r_1(x_1) = 0$$
$$r_2(x_1, x_2, x_3) = 0$$
$$r_3(x_1, x_2, x_3) = 0$$
$$r_4(x_1, x_2, x_3, x_4) = 0$$

That is, the first equation contains just the first unknown, x_1, whereas the next two equations contain the first three unknowns, x_1, x_2, and x_3, and only the last, fourth, equation contains all the four unknowns.

This situation suggests an obvious plan to deal with the proposed system of equations: We begin with x_1 which we compute from the first equation. Having obtained the value of x_1, we observe that the next two equations form a system from which we can determine the next two unknowns, x_2 and x_3. Having so obtained x_1, x_2, and x_3, we use the last, fourth, equation to compute the last unknown x_4.

(2) Let us now realize that the system of equations considered expresses the *condition* of a problem. This condition is split into four parts and each single equation represents a part (or clause, or proviso) of the full condition: the equation expresses that the unknowns involved are connected with each other and with the data by just such a relation as the corresponding part, or clause, of the condition prescribes. And so the condition has a peculiar feature: not all of its clauses involve all the unknowns. Our notation clearly shows which clause involves which unknowns.

Of course, the condition can be split into clauses in just this peculiar manner (with each clause involving just the indicated particular combination of unknowns) even if we have not yet translated those clauses into equations, or even if we are not able to translate those clauses into equations. We may suspect that the *plan* sketched above, under (1), for a system of equations may *remain valid* in some sense for a system of clauses even if those clauses are not expressed, or are not expressible, algebraically.

This remark opens a broad vista of new possibilities.

(3) In order to see these possibilities more clearly we have to reinterpret our notation.

Till now we have interpreted the symbol $r(x_1, x_2, \ldots, x_n)$ in the usual way as an algebraic expression in (or a polynomial in, or a function of) the unknown (variable) numbers $x_1, x_2, \ldots, x_n$. And so we have interpreted

$$r(x_1, x_2, \ldots, x_n) = 0$$

as an (algebraic) equation linking the unknowns $x_1, x_2, \ldots, x_n$. If we deal with a problem in which $x_1, x_2, \ldots, x_n$ are unknowns, such an equation expresses a part of the condition (a clause or a proviso of the condition), that is, a relation between the unknowns $x_1, x_2, \ldots, x_n$ and the data required by the condition.

We do not repudiate this interpretation but we do extend it: Even if the clause is not translated into an equation, or even if $x_1, x_2, \ldots, x_n$ are not unknown numbers but unknown things of any kind, the symbolic equation

$$r(x_1, x_2, \ldots, x_n) = 0$$

should express a *relation, required by the condition of the problem, which involves the indicated unknowns* $x_1, x_2, \ldots, x_n$. We may also say that such a symbolic equation expresses a part of the condition (a clause, proviso, stipulation, or requirement imposed by the condition).

We need a few examples to understand properly this extended scope of the notation, and still more examples to convince ourselves that this extension is useful.

(4) The notation that we have just introduced can be suitably illustrated by crossword puzzles.   Let us look at a (miniature) example.

Across	Down
Across	*Down*
1. German mathematician	1. Do not write dagre
2. Do not write beaus	5. This is boring as
3. Swiss mathematician	6. Reset reset.

In a crossword puzzle the unknowns are words.   Let $x_1, x_2, \ldots, x_6$ stand for the six unknown words of our puzzle.   Both words x_1 and x_4 have their initial letter in the same square numbered 1, but x_1 should be written horizontally across and x_4 vertically downward; if $n = 2, 3, 5$, or 6, x_n stands for the word the initial letter of which should be written in the square numbered n.   If we spell out pedantically the conditions implied by the square diagram that contains black and white, numbered and unnumbered, smaller squares, we have a system of 21 conditions.

There are the six most conspicuous conditions expressed by the "clues." Let us represent them by

$$r_1(x_1) = 0, \quad r_2(x_2) = 0, \quad \ldots, \quad r_6(x_6) = 0$$

Thus, the symbolic equation $r_1(x_1) = 0$ represents the condition that the word x_1 is the name (we hope the last name) of a German mathematician;

$r_4(x_4) = 0$ expresses the import of the (for the moment, rather cryptic) sentence, "Do not write dagre," and so on.

There are the six conditions, visible from the diagram, concerned with the lengths of the six unknown words:

$$r_7(x_1) = 0, \quad r_8(x_2) = 0, \quad \ldots, \quad r_{12}(x_6) = 0$$

For instance, $r_7(x_1) = 0$ prescribes the length of the word x_1. The meaning in our case is that each of the words, $x_1, x_2, \ldots, x_6$ should be a five-letter word.

The diagram shows which word crosses which other and where, and so implies nine conditions:

$$\begin{aligned}
r_{13}(x_1, x_4) &= 0, & r_{14}(x_1, x_5) &= 0, & r_{15}(x_1, x_6) &= 0 \\
r_{16}(x_2, x_4) &= 0, & r_{17}(x_2, x_5) &= 0, & r_{18}(x_2, x_6) &= 0 \\
r_{19}(x_3, x_4) &= 0, & r_{20}(x_3, x_5) &= 0, & r_{21}(x_3, x_6) &= 0
\end{aligned}$$

For example, $r_{14}(x_1, x_5) = 0$ stipulates that the third letter of the word x_1 is the initial of the word x_5, and so on.

Now, we have listed all conditions; their number is $6 + 6 + 9 = 21$.

(5) In general, if the problem involves n unknowns $x_1, x_2, \ldots, x_n$ and we split the condition into l different parts (requirements, stipulations, clauses, provisos) we have a system of l relations connecting n unknowns which we may express by a system of l symbolic equations connecting those n unknowns as follows:

$$\begin{aligned}
r_1(x_1, x_2, \ldots, x_n) &= 0 \\
r_2(x_1, x_2, \ldots, x_n) &= 0 \\
\cdot \quad \cdot \quad \cdot \quad \cdot \quad \cdot \quad \cdot \quad \cdot \\
r_l(x_1, x_2, \ldots, x_n) &= 0
\end{aligned}$$

In chapter 2 we were concerned with the particular case in which the unknowns $x_1, x_2, \ldots, x_n$ are unknown numbers, the equations not merely symbolic, but actual algebraic equations, and $l = n$. In the present chapter we shall often be concerned with special situations, such as that discussed under (1) and (2), in which not all the clauses involve all the unknowns.

(6) It can happen that two problems are expressed by the same system of symbolic equations. Such problems may deal with very different matters, but they have something in common: They are similar to each other in some (rather abstract) respect, we may put them into the same class. In this way we obtain a new, more refined, classification of problems (of problems *to find*). Has this classification some interest for our study? If two problems are expressed by the same system of symbolic equations, is there a procedure of solution applicable to both?

This is a good question, I think. Taken in full generality it may be not very useful, but it helps to understand the special situations which we are going to discuss.

6.2. Wider scope of the pattern of two loci

In the foregoing sect. 6.1 we have outlined a very general picture. How do our former observations fit into this picture? How does the very first pattern that we have discerned fit into it?

(1) We may give the answer in a more striking form if we adapt our terminology.

In dealing with geometric constructions, we considered "loci." Such a locus is really just a set of points. In what follows we shall call a set a *locus* if it intervenes in the solution of a problem in a certain characteristic manner which will be indicated by the following examples. As the term "set," see ex. 1.51, has already so many synonyms (class, aggregate, collection, category) it may seem wanton to add one more. Yet the term "locus" may remind us of our experience with certain elementary geometric problems and so it may suggest, by analogy, useful steps when we are dealing with other, perhaps more difficult, problems.

(2) *Two loci for a point in the plane.* Let us return to the very first example that we have discussed: *Construct a triangle being given its three sides.*

Let us look back at the familiar solution (sect. 1.2). By laying down one side, say a, we locate two vertices, B and C, of the required triangle. Just one more vertex remains to be found; call this third vertex, still unknown at this stage of our work, x. The condition requires two things of this point x:

(r_1) the point x is at the given distance b from the given vertex C,
(r_2) the point x is at the given distance c from the given vertex B.

Using the notation introduced in sect. 6.1, we write these two requirements, (r_1) and (r_2), as two symbolic equations:

$$r_1(x) = 0$$
$$r_2(x) = 0$$

The points x satisfying the first requirement (r_1) (the first of the two symbolic equations) fill the periphery of a circle (with center C and radius b). This circular line forms the *set*, or *locus*, of all points complying with the requirement (r_1). The locus of the points satisfying the second requirement (r_2) (the second symbolic equation) is another circular line. Now, the point x that solves the proposed problem about the triangle has to satisfy both requirements, it must belong to both loci. Therefore, the intersection of these two loci is the set of the solutions of the proposed problem. This set contains two points: there are two solutions, two triangles symmetric to each other with respect to the side *BC*.

(3) *Three loci for a point in space.* We consider the following problem of solid geometry, which is analogous to the simple problem of plane geometry that we have just discussed, under (2): *Find a tetrahedron being given its six edges.*

By the procedure that we have just recalled, under (2), we construct the base of the tetrahedron, a triangle, from the three edges that are required to surround it. Laying down the base, we locate three vertices of the tetrahedron, say *A*, *B*, and *C*. Just one more vertex remains to be found; call this fourth vertex, still unknown at this stage of our work, *x*, and call the given distances from the three already located vertices *a*, *b*, and *c*, respectively. The condition requires three things of the point *x*:

(r_1), *x* is at the distance *a* from the point *A*,
(r_2), *x* is at the distance *b* from the point *B*,
(r_3), *x* is at the distance *c* from the point *C*.

Using the notation introduced in sect. 6.1, we write these three requirements, (r_1), (r_2), and (r_3), as three symbolic equations

$$r_1(x) = 0$$
$$r_2(x) = 0$$
$$r_3(x) = 0$$

The points *x* satisfying the first requirement (r_1) (the first symbolic equation) fill the surface of a sphere (with center *A* and radius *a*). This spherical surface forms the set, or locus, of all points complying with the first requirement (r_1). To each of the other two requirements there corresponds a spherical surface, the locus of the points *x* satisfying that requirement. Now the point *x* that solves the proposed problem about the tetrahedron has to satisfy all three requirements, it must belong to all three loci. Therefore, the intersection of these three loci (three spheres) is the set of the solutions of the proposed problem. This set contains two points: there are two solutions, two tetrahedra symmetric to each other with respect to the plane of the triangle *ABC*.

(4) *Loci for a general object.* The examples discussed under (2) and (3) may remind us of several other problems that we have solved in chapter 1 following the same pattern. Behind these examples we may perceive a general situation.

The unknown of a problem is x. The condition of the problem is split into l clauses which we express by a system of l symbolic equations:

$$r_1(x) = 0, \quad r_2(x) = 0, \quad \ldots, \quad r_l(x) = 0$$

Those objects x that satisfy the first clause, expressed by the first symbolic equation, form a certain set which we call the first locus. The objects satisfying the second clause form the second locus, . . . the objects satisfying the last clause form the lth locus. The object x that solves the proposed problem must satisfy the full condition, that is, all the l clauses of the condition, and so it must belong to all l loci. On the other hand, any object x that does belong simultaneously to all l loci satisfies all the l clauses, and so the full condition, and is, therefore, a solution of the proposed problem. In short, the *intersection* of those l loci constitutes the set of solutions, that is, the set of all objects satisfying the condition of the proposed problem.

This suggests a vast generalization of the pattern of two loci, a scheme that could work in an inexhaustible variety of cases, could solve almost any problem: first, split the condition into appropriate clauses, then form the loci corresponding to the various clauses, finally find the solution by taking the intersection of those loci. Before judging this vast scheme, let us get down to concrete cases.

(5) *Two loci for a straight line.* Construct a triangle being given r, h_a, and α.

The reader should remember the notation used in chapter 1: r stands for the radius of the inscribed circle, h_a for the height perpendicular to the side a, and α for the angle opposite the side a.

The problem is not too easy, but certain initial steps are rather obvious. *Could you solve a part of the problem?* We can easily draw a part of the required figure: a circle with radius r and two tangents to it that include the angle α. (Observe that the two radii drawn to the two points of contact include the angle $180° - \alpha$.) The vertex of this angle α will be the vertex A of the required triangle. The problem is reduced to the construction of the (infinite) straight line of which the side opposite A is a segment. This line, say x, is our new unknown, given that part of the figure that we have already drawn.

The condition for the line x consists of two clauses:

(r_1), x is tangent to the circle with radius r already constructed,
(r_2), x is at the given distance h_a from the given point A.

The first locus for x is the set of the tangents of the given circle with radius r.

The second locus for x is again the set of the tangents of a circle the center of which is A and the radius h_a.

The intersection of these two loci consists of the common tangents of the two circles.   We can construct these tangents; see sect. 1.6(1) and ex. 1.32.

(In fact, only the exterior common tangents solve the problem as stated; the interior common tangents, which may not exist, would render the circle with radius r an escribed circle.)

Regarding the common tangents of two circles as the *intersection of two loci for straight lines* is a useful idea; it is even more useful if we include in it similar cases, especially the extreme case in which one of the circles degenerates into a point.

(6) *Three loci for a solid.*   Design a "multipurpose plug" that fits exactly into three different holes, circular, square, and triangular.

See Fig. 6.1; the diameter of the circle, the side of the square, the base and the altitude of the isosceles triangle are equal to each other.

In geometric terms, three orthogonal projections of the required solid should coincide with the three given shapes.   We assume (in fact, this assumption narrows down the question) that the directions of the three projections are perpendicular to each other.   Our unknown is a solid, say x, and the condition of our problem consists of three clauses:

(r_1) the projection of x onto the floor is a circle,
(r_2) the projection of x onto the front wall is a square,
(r_3) the projection of x onto the side wall is an isosceles triangle.

It is understood that the solid x is placed into a room having the usual shape of a rectangular parallelepiped, that the projections are orthogonal, and that the measurements of the three shapes in Fig. 6.1 are related as has been explained.

Let us examine the first locus, that is, the set of solids satisfying the requirement (r_1).   The given circle is placed on the floor.   Consider any infinite vertical straight line passing through the area of this circle and call it a "fiber."   These fibers fill an infinite circular cylinder of which the given circle is a cross section.   A solid x satisfies the first requirement (r_1)

Fig. 6.1. Three holes for the multipurpose plug.

if it is contained in the cylinder and contains at least one point of each fiber. The set of all such solids is the first locus.

As the first locus is connected with the infinite vertical cylinder so are the two other loci connected with two infinite horizontal prisms. The cross section of the prism corresponding to (r_2) is a square. If this prism lies in the north-south direction, the prism corresponding to (r_3) which has a triangular cross section lies in the east-west direction.

Any solid x that belongs to all three loci solves the problem, is a "multipurpose plug." The most extensive solid of this kind is the intersection of the three infinite figures, of the cylinder and the two prisms; it is sketched in Fig. 6.2.

(Why is it the most extensive? Describe the various parts of its surface. Describe some other solids that solve the problem.)

(7) *Two loci for a word.* In a crossword puzzle that allows puns and anagrams we find the following clue:

"This form of rash aye is no proof (7 letters)."

This is a vicious little sentence; it almost makes sense: "If you say Yes so rashly, it does not prove a thing." Yet we suspect that some vague echo of a sense was put into the clue just to lead us astray. There may be a better lead: the phrase "form of" may mean "anagram of." And so we may try to interpret the clue as follows.

The unknown x is a word. The condition consists of two parts:

(r_1), x is an anagram of (has the same seven letters as) RASH AYE;

(r_2), "x is no proof" is a meaningful (probably usual) phrase.

Let us examine this interpretation of the problem. The condition is neatly split into two clauses: (r_1) is concerned with the spelling of the word, (r_2) with its meaning. To each clause corresponds a "locus"—but these loci are less "manageable" than in the foregoing cases.

The first locus is quite clear in itself. We can order the seven letters

A A E Y H R S

in 2520 different ways (the reader need not examine now the derivation of

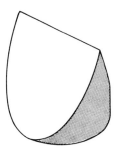

Fig. 6.2. The best multipurpose plug.

this number which equals, in fact, $7!/2!$). If it were absolutely necessary, we could write down the 2520 different arrangements of the 7 given letters without repetition or omission and so exhaust the possibilities left open by clause (r_1), that is, describe or construct completely the corresponding locus. This, however, would be boring and wasteful (many of the arrangements would have combinations of vowels and consonants never arising in English). Moreover, such a mechanical exhaustion of all cases was not intended, it is not in the spirit of the game. And so the locus corresponding to clause (r_1) is, if not in principle, but in practice, inexhaustible, unmanageable.

The locus corresponding to clause (r_2) is not only inexhaustible but somewhat hazy. An English word x is given; does the phrase "x is no proof" make sense? Is it a usual phrase? In many cases, the answer is debatable.

And so, for different reasons, neither of the two loci is manageable, neither can be conveniently described, surveyed, or constructed. And, of course, we have no clear procedure to construct the intersection of the two loci. Still, it may be helpful to realize that the condition has two different clauses and that the required word has to satisfy both. Focusing now one clause of the condition, then the other, thinking of words which almost fulfill one clause, or the other, a stab in this direction, then one in the other—eventually, our memory, our store of words and phrases may be sufficiently stirred, and the desired word may pop up.

(We have insisted on the circumstance that neither of the two clauses, (r_1) and (r_2), is manageable—this point is useful in assessing the general scheme we are considering. In fact, however, one of the two clauses is somewhat more manageable than the other—this point may be useful in solving the little riddle at hand.)

6.3. The clause to begin with

In the foregoing section we have discussed problems of various kinds and solved them following the same pattern which we may call the "pattern of l loci." Yet we did not solve the last problem, in sect. 6.2(7). What was the difficulty? We succeeded in splitting the condition into clauses quite neatly, but we failed to manage the loci corresponding to these clauses: we could not exhaust, could not describe conveniently those loci and so we could not form their intersection.

There are cases in which we have this difficulty but not in its most formidable form, and we may be able to handle such cases.

(1) *Two loci for a word.* In a crossword puzzle that allows puns and anagrams we find the following clue:
"Flat both ways (5 letters)".

After some trials we may be led to the following interpretation: The unknown x is a word. The condition consists of two clauses:

(r_1), x means "flat"

(r_2), x is a word having 5 letters which, read backward, still has the same meaning "flat."

With which clause should we begin? There is a difference. To manage the clause (r_2) efficiently, you should have in your head a list of all five-letter words that can be read backward with some meaning. Now, very few of us have such a list. But most of us can remember words that have more or less the same meaning as "flat"; we have to examine them as they emerge whether they also fulfill the clause (r_2). Here are some such words:

even, smooth, unbroken—plain, dull—horizontal—of course, LEVEL![1]

(2) Let us try to disentangle the essential feature of the foregoing procedure.

The clause (r_1) selects from the vast range of all words a small set of words, one among which is the solution. The clause (r_2) does the same, but there is a difference: the selection is easier in one case than in the other, we can handle (r_1) more efficiently than (r_2). We have used the more manageable clause for a first selection and the less manageable clause for a subsequent second selection. It is more necessary to be efficient in the first selection: we select elements the first time from the immense reservoir of all words, the second time from the very much more restricted first locus, obtained by the first selection.

The moral is simple: To each clause corresponds a locus. *Begin with the clause, for which the locus can be more fully, or more efficiently, formed.* Doing so, you may avoid forming the loci corresponding to the other clauses: you use those other clauses in selecting elements from the first locus.

(3) *Two loci for a tripartite unknown.* How old is the captain, how many children has he, and how long is his boat? Given the product 32118 of the three desired numbers (integers). The length of the boat is given in feet (is several feet), the captain has both sons and daughters, he has more years than children, but he is not yet one hundred years old.

This puzzle demands to find three numbers,

$$x, \qquad y, \qquad z$$

which represent the captain's

        number of children,   age,   length of boat

---

[1] HSI, Decomposing and recombining 8, pp. 83–84, contains a very similar example and anticipates the essential idea of the present section.

respectively. It will be advantageous to conceive the problem thus: We have but one unknown; this unknown, however, is not a number but a tripartite unknown, a triplet (x, y, z) of numbers.

It is very important to split the condition that is expressed by the statement of the problem into appropriate clauses. This needs careful consideration of details and considerable regrouping. After several trials (which we skip to save space) we may arrive at the following two clauses:

(r_1), x, y, and z are positive integers different from 1 and such that

$$xyz = 32118$$

(r_2) $$4 \leqq x < y < 100$$

With which one of the two clauses should we begin? Of course, with (r_1) which leaves only a finite number of possibilities, whereas (r_2), which does not restrict z at all, leaves an infinite number.

Therefore, we examine (r_1). Now, 32118 is divisible by 6, and so we easily decompose it into prime factors:

$$32118 = 2 \times 3 \times 53 \times 101$$

For a decomposition into three factors we have to combine two of the four primes. Therefore, there are only six different ways to decompose the number 32118 into a product of three factors all different from 1:

$$6 \times \quad 53 \times 101$$
$$3 \times 101 \times 106$$
$$3 \times \quad 53 \times 202$$
$$2 \times 101 \times 159$$
$$2 \times \quad 53 \times 303$$
$$2 \times \quad 3 \times 5353$$

Of these six possibilities, the remaining requirement (r_2) rejects all except the first one, and so we obtain

$$x = 6, \quad y = 53, \quad z = 101$$

The captain has 6 children, is 53 years old, and the length of his boat is 101 feet.

The essential idea of the solution of this simple puzzle is often applicable, also in more complicated cases: split off from the full condition a "major" clause that leaves open only a small number of possibilities, and choose between these possibilities by using the remaining "minor" part of the condition.[2]

†(4) *Two loci for a function.* There is a very important type of mathematical problem, of daily use in physics and engineering, the condition of

---

[2] Cf. ex. 6.12 to 6.17.

which is naturally split into two clauses: to determine a function by a differential equation and initial, or boundary, conditions. Here is a simple example: the unknown x is a function of the independent variable t; it is required to satisfy

(r_1) the differential equation $\dfrac{d^2x}{dt^2} = f(x, t)$, where $f(x, t)$ is a given

function,

(r_2) the initial conditions $x = 1, \dfrac{dx}{dt} = 0$ for $t = 0$.

Should we begin with the differential equation or with the initial conditions? That depends on the nature of the given function $f(x, t)$.

First case. Take $f(x, t) = -x$, so that the proposed differential equation is

$$\frac{d^2x}{dt^2} = -x$$

This differential equation belongs to those few privileged types of which we can exhibit the "general integral" explicitly. In fact, the most general function satisfying the differential equation is

$$x = A \cos t + B \sin t$$

where A and B are arbitrary constants (constants of integration). Thus we have obtained the "locus" corresponding to the clause (r_1).

We proceed now to the clause (r_2) which we use to pick out the solution from the first locus that we have just obtained: setting $t = 0$ in the expressions for x and $\dfrac{dx}{dt}$, we find from the initial conditions that

$$A = 1, \qquad B = 0, \qquad x = \cos t$$

Second case. We examine the differential equation, but do not succeed in finding its general integral (or any of its integrals) and decide that we shall not make any further efforts in this direction. What should we do next? With which one of the two clauses, (r_1) and (r_2), should we now begin?

In this situation we may use (r_2) first: we set up x as a power series in t, the initial coefficients of which are determined by the initial conditions whereas the remaining coefficients $u_2, u_3, u_4, \ldots$ appear, at this stage of our work, undetermined (they are, in fact, our unknowns, see ex. 3.81):

$$x = 1 + u_2 t^2 + u_3 t^3 + u_4 t^4 + \cdots$$

Thus, in a sense, the locus corresponding to (r_2) has been obtained. We now proceed to (r_1), the first clause, to determine the remaining coefficients

u_2, u_3,... from the differential equation (by recursion, if possible; see again ex. 3.81).

Observe that, in any case, the differential equation is more "selective" (narrows down the choice of the function much more) than the initial condition. Thus, the proposed (r_2) determines only two coefficients of the power series; the differential equation (the condition (r_1)) has to determine the remaining infinite sequence. This shows that the more selective clause is not always the best to begin with.

6.4. Wider scope for recursion

In the foregoing section we have observed an important difference between clauses and clauses: there may be reasons, and even strong reasons, to begin the work rather with one clause than with the other. It is true, there was a limitation: we have considered the case of one unknown. (This limitation is not really restrictive; see the indication in sect. 5.5.) Let us now consider the case of several unknowns.

(1) There is an important general situation which is suggested by several examples considered in chapter 3. There are n unknowns $x_1, x_2, x_3, \ldots, x_n$ which satisfy n conditions of the following form:

$$r_1(x_1) = 0$$
$$r_2(x_1, x_2) = 0$$
$$r_3(x_1, x_2, x_3) = 0$$
$$\cdot \quad \cdot \quad \cdot \quad \cdot \quad \cdot \quad \cdot \quad \cdot$$
$$r_n(x_1, x_2, x_3, \ldots, x_n) = 0$$

This particular system of n relations suggests not only where we should begin, but also how we should go on. In fact, it suggests a full plan of campaign: Begin with x_1, which you should determine from the first relation. Having obtained x_1, determine x_2 from the second relation. Having obtained x_1 and x_2, determine x_3 from the third relation, and so on: determine the unknowns $x_1, x_2, \ldots, x_n$ one at a time, in the order in which they are numbered, using the values of those already obtained in determining the next one. This plan works well if the kth relation is an equation

$$r_k(x_1, x_2, \ldots, x_{k-1}, x_k) = 0$$

from which we can express x_k in terms of $x_1, x_2, \ldots, x_{k-1}$, for $k = 1, 2, 3, \ldots, n$. The situation is particularly favorable if the kth equation is linear with respect to x_k (the coefficient of which should not vanish, of course).

This is the pattern of *recursion*: we determine x_k by recurring, or going back, to the previously obtained $x_1, x_2, \ldots, x_{k-1}$.

Following this pattern we simply proceed step by step, beginning with x_1, tackling x_2 after x_1, x_3 after x_2, and so on, which seems to be the most obvious, the most natural thing to do. At each step we refer to information accumulated by the foregoing steps—and this is perhaps the most significant feature of the pattern. We shall see the point more clearly after a few examples.

(2) In sect. 2.5(3) we obtained a system of 7 equations for 7 unknowns. Let us relabel the unknowns as follows:

$$D = x_7$$
$$a = x_4 \qquad b = x_5 \qquad c = x_6$$
$$p = x_1 \qquad q = x_2 \qquad r = x_3$$

And let us rewrite the system of equations, expressing precisely which unknowns are linked by each equation, but disregarding other details, and numbering the equations so that the order in which they should be treated is clearly visible.

Thus we obtain the following system of relations:

$$r_1(x_2, x_3) = 0$$
$$r_2(x_3, x_1) = 0$$
$$r_3(x_1, x_2) = 0$$
$$r_4(x_2, x_3, x_4) = 0$$
$$r_5(x_3, x_1, x_5) = 0$$
$$r_6(x_1, x_2, x_6) = 0$$
$$r_7(x_4, x_5, x_6, x_7) = 0$$

So written, the system renders the following plan obvious: Let us separate the first three relations from the rest. They contain only the first three unknowns x_1, x_2, x_3 and may be regarded as a system of three equations for these three unknowns. (In fact, we can easily express $x_1 = p$, $x_2 = q$, $x_3 = r$ from the system of the three equations given in sect. 2.5(3) that are indicated here by the first three relations.) Once the first three unknowns x_1, x_2, x_3 have been found, the system "becomes recursive": First, we obtain x_4, x_5, x_6, each unknown from the correspondingly numbered relation. (In fact, the order in which we treat these three unknowns does not matter.) Having found x_4, x_5, x_6, we use the last relation to obtain x_7 (which is the principal unknown in the original problem, see sect. 2.5(3); the others are only auxiliary unknowns).

The reader should compare the system just discussed with the system considered in sect. 6.1(1).

(3) *Solve the equation*

$$(he)^2 = she$$

Of course, *he* and *she* are ordinary numbers (positive integers) written in the ordinary decimal notation, the one a two-digit number, the other a three-digit number, and h, e, and s are digits. We may restate the problem quite fussily: find h, e, and s satisfying

$$(10h + e)^2 = 100s + 10h + e$$

where h, e, and s are integers, $1 \leq h \leq 9$, $0 \leq e \leq 9$, $1 \leq s \leq 9$.

This little puzzle is not difficult. If the reader has solved it by himself, he will be in a better position to appreciate the following scheme. In the initial phase we shall examine just one unknown. In the next phase we shall bring in one more and consider the two unknowns jointly. Only in the last phase shall we deal with all three unknowns.

Phase (e). We begin with e, since there is a *requirement for e alone*: the last decimal of e^2 must be e. We list the squares of the ten digits,

$$0, \quad 1, \quad 4, \quad 9, \quad 16, \quad \mathbf{25}, \quad \mathbf{36}, \quad 49, \quad 64, \quad 81$$

and find that there are only four out of ten that satisfy the requirement, and so

$$e = 0 \quad \text{or} \quad 1 \quad \text{or} \quad 5 \quad \text{or} \quad 6$$

Phase (e, h). There is a requirement that involves just two out of the three digits, e and h:

$$100 \leq (he)^2 < 1000$$

from which we easily conclude that

$$10 \leq he \leq 31$$

Combining this information with the result of the foregoing consideration under *(e)*, we find that the two-digit number *he* must be one of the following ten numbers:

$$
\begin{array}{cccc}
10, & 11, & 15, & 16, \\
20, & 21, & 25, & 26, \\
30, & 31 &
\end{array}
$$

Phase (e, h, s). Now we list the squares of the ten numbers just obtained

$$
\begin{array}{cccc}
100, & 121, & 225, & 256 \\
400, & 441, & \mathbf{625}, & 676 \\
900, & 961 &
\end{array}
$$

and we find just one that satisfies the full condition. Hence

$$e = 5, \quad h = 2, \quad s = 6$$

$$(25)^2 = 625$$

(4) In the foregoing subsection (3) we have split the condition of the

proposed problem into three clauses which we may represent (using the notation introduced in sect. 6.1) by a system of three symbolic equations

$$r_1(e) = 0$$
$$r_2(e, h) = 0$$
$$r_3(e, h, s) = 0$$

Let us compare this system of three clauses with the following system of three linear equations:

$$a_1 x_1 = b_1$$
$$a_2 x_1 + a_3 x_2 = b_2$$
$$a_4 x_1 + a_5 x_2 + a_6 x_3 = b_3$$

x_1, x_2, x_3 are the unknowns, a_1, $a_2, \ldots, a_6$, $b_1, \ldots, b_3$ are given numbers, a_1, a_3, and a_6 are supposed to be different from 0.

The similarity of these two systems is more obvious than their difference; let us compare them carefully.

Let us look first at the system of three linear equations for x_1, x_2, and x_3. The first equation determines the first unknown x_1 completely: the later equations will in no way influence or modify the value of x_1 obtained from the first. Based on this value of x_1, the second equation completely determines the second unknown x_2.

The system of the three clauses into which we have split the condition for the unknowns e, h, and s is formally similar to, but materially different from, the system of three equations for x_1, x_2, x_3. The first clause does not completely determine the first unknown e; it just narrows down the choice for e; it yields (this is the most appropriate expression) a *locus* for e. Similarly, the second clause does not completely determine the second unknown h; it yields a locus for the couple of two unknowns (e, h). Only the last clause achieves a definitive determination: it picks out from the locus previously established the only triplet (e, h, s) that satisfies the full condition.

6.5. Gradual conquest of the unknown

Considering n numerical unknowns x_1, $x_2, \ldots, x_n$, we may regard them as the successive components of one multipartite unknown x (see sect. 5.5). Let us so view the n unknowns which we determine successively from a recursive system of equations such as we have considered in sect. 6.4(1). The recursive procedure of solution unveils our multipartite unknown gradually, step by step. At first we obtain little information about the unknown, the value of just one component, x_1. Yet, using this initial information to advantage, we obtain more: we add the knowledge of the second component x_2 to that of the first. At each stage of our work we add the knowledge of one more component to our previously acquired knowledge, at each stage we use the information already obtained to obtain additional information. We conquer an empire province by province, using at each stage the provinces already won as *base of operations* to win the next province.

We have seen cases in which this procedure is more or less modified. The provinces may not be conquered exactly one at a time, but the empire builder sometimes takes a bigger bite, two or three provinces at the same time; cf. sect. 6.4(2) and sect. 6.1(1). Or a province is not conquered fully at one stroke; first one, then another province is made partially dependent, and a final successful move acquires all at once; cf. sect. 6.4(3).

We may have met with still other variations of the procedure in our past experience. We certainly had opportunity to be impressed with the peculiar expanding pattern of the work for the solution; cf. sect. 2.7. If the unknown has many components (as in a crossword puzzle), we may advance along several lines simultaneously: we need not thread all our beads on one string, but may use several strings. Yet the essential thing is to *use the information already gathered as a base of operations to gather further information.* Perhaps all rational procedures of problem solving and learning are recursive in this sense.

Examples and Comments on Chapter 6

6.1. *A condition with many clauses.* In a *magic square* with n rows, n^2 numbers are so arranged that the sum of the numbers in each of the n rows, in each of the n columns and in each of the two diagonals is the same; this sum is called the "constant" of the magic square. The simplest and best-known magic square has $n = 3$ rows and is filled with the first nine natural numbers $1, 2, \ldots, 9$. Let us state in minute detail the problem that requires us to find this simple magic square.

What is the unknown? There are nine unknowns; let x_{ik} denote the desired number in the ith row and the kth column; $i, k = 1, 2, 3$.

What is the condition? The condition has four different kinds of clauses:

(1) x_{ik} is an integer
(2) $1 \leqq x_{ik} \leqq 9$
(3) $x_{ik} \neq x_{jl}$ unless $i = j$ and $k = l$
(4) $x_{i1} + x_{i2} + x_{i3} = x_{11} + x_{22} + x_{33}$ for $i = 1, 2, 3$
    $x_{1k} + x_{2k} + x_{3k} = x_{11} + x_{22} + x_{33}$ for $k = 1, 2, 3$
    $x_{13} + x_{22} + x_{31} = x_{11} + x_{22} + x_{33}$

State the number of clauses of each kind and the total number of clauses. Which shape have these clauses in the notation of sect. 6.1(3)?

6.2. By introducing a multipartite unknown reduce the general system considered in sect. 6.1(5) to the (apparently more particular) system considered in sect. 6.2(4).

6.3. By introducing a multipartite unknown reduce the system considered in sect. 6.1(1) to a particular case of the system considered in sect. 6.4(1).

6.4. Reduce the system considered in sect. 6.4(2) in the manner of ex. 6.3.

6.5. Devise a plan to solve the system

$$r_1(x_1, x_2, x_3) = 0$$
$$r_2(x_1, x_2, x_3) = 0$$
$$r_3(x_1, x_2, x_3) = 0$$
$$r_4(x_1, x_2, x_3, x_4) = 0$$
$$r_5(x_1, x_2, x_3, x_5) = 0$$
$$r_6(x_1, x_2, x_3, x_6) = 0$$
$$r_7(x_1, x_2, x_3, x_4, x_5, x_6, x_7) = 0$$

6.6. The system of relations

$$r_1(x_1) = 0$$
$$r_2(x_1, x_2) = 0$$
$$r_3(x_2, x_3) = 0$$
$$r_4(x_3, x_4) = 0$$
$$.\quad.\quad.\quad.\quad.\quad.$$
$$r_n(x_{n-1}, x_n) = 0$$

is a particularly interesting particular case of a system considered in the text: of which one?

Have you seen it before? Where have you had opportunity to compare two systems analogously related to each other?

6.7. Through a given interior point of a circle construct a chord of given length.

Classify this problem.

6.8. Two straight lines, a and b, and a point C are given in position; moreover, a length l is given. Draw a straight line x through the point C so that the perimeter of the triangle formed by the lines a, b, and x is of length l.

Classify this problem.

6.9. *Keep only a part of the condition.* Of the two clauses of the problem considered in sect. 6.2(7), (r_1) is somewhat more manageable: in trying to satisfy this requirement, we can map out some sort of a plan. To find an anagram of a given set of letters, such as RASH AYE, we have to find a word that has only letters of this set and has all of them. A procedure that may help is the following: drop the last part of the condition "and has all of them" and try to find words, or usual word parts, word endings, formed with the letters of the given set. Short words of this kind are easy to form, and proceeding to longer ones we may hope to arrive eventually at the desired anagram.   In our case, we may hit upon the following:

> ASH,   YES,   SAY,   SHY,   RYE,   EAR
> HEAR,   HARE,   AREA
> SHARE
> RE- (beginning)
> -ER,   -AY,   -EY (endings).

To solve the problem of sect. 6.2(7), we look at these bits of words, having in mind not only the anagram or clause (r_1), but also clause (r_2).   Some of the bits may combine into a full anagram—but SHY AREA is not an acceptable solution.

6.10. *The thread of Ariadne.*   The daughter of King Minos, Ariadne, fell in love with Theseus and gave him a clue of thread which he unwound when entering the Labyrinth and found his way out of its mazes by following back the thread.

Did some prehistoric heuristic genius contribute to the formation of this myth?   It suggests so strikingly the nature of certain problems.

In trying to solve a problem, we often run into the difficulty that we see no way to proceed farther from the last point that we have attained.   The Labyrinth suggests another kind of problem in which there are many ways proceeding from each point attained, but the difficulty is to choose between them. To master such a problem (or to present its solution when we have succeeded in mastering it) we should try to treat the various topics involved *successively,* in the most appropriate, the most *economical* order: whenever an alternative presents itself, we should choose *the next topic so that we can derive the maximum help from our foregoing work.*   The "most appropriate choice at the crossroads" is strongly suggested by the "thread of Ariadne" which was, by the way, one of Leibnitz's pet expressions.

Problems involving several unknowns, several interrelated tasks and conditions, are often of such a labyrinthine nature; crossword puzzles and constructions of complex geometric figures may yield good illustrations.   Having to solve such a problem, we have a choice at each stage: to which task (to which word, to which part of the figure) should we turn next?   At first, we should look for the weakest spot, for the clause to begin with, for the most accessible word of the puzzle, or the most easily constructible part of the figure.   Having found a first word or having constructed a first bit of the figure, we should carefully select our second task: that word (or that part of the figure) different from the first to find which the first word (or part) already found offers the most help.   And so on, we should always try to select our next task, the next unknown to find, so that we get the maximum help from the unknowns previously found.   (This spells out an idea already voiced in sect. 6.5.)

There follow a few problems which give an opportunity to the reader to try out the preceding advice.

6.11. Find the magic square with three rows minutely described in ex. 6.1. (You may know one solution, but you should find all solutions. The order in which you examine the various unknowns is very important. Especially, try to spot those unknowns whose values are uniquely determined and begin with them.)

6.12. Multiplication by 9 reverses a four-digit number (produces a four-digit number with the same digits in reverse order). What is the number? (Which part of the condition will you use first?)

6.13. Find the digits a, b, c, and d, being given that

$$ab \times ba = cdc$$

It is assumed that the digits a and b of the two-digit number ab (that is $10a + b$) are different.

6.14. A triangle has six "parts": three sides and three angles. Is it possible to find two such *noncongruent* triangles that five parts of the first are identical with five parts of the second? (I did *not* say that those five identical parts are *corresponding* parts.)

6.15. Al, Bill, and Chris planned a big picnic. Each boy spent 9 dollars. Each bought sandwiches, ice cream, and soda pop. For each of these items the boys spent jointly 9 dollars, although each boy split his money differently and no boy paid the same amount of money for two different items. The greatest single expense was what Al paid for ice cream; Bill spent twice as much for sandwiches as for ice cream. How much did Chris pay for soda pop? (All amounts are in round dollars.)

6.16. In preparation for Hallowe'en, three married couples, the Browns, the Joneses, and the Smiths, bought little presents for the neighborhood youngsters. Each bought as many identical presents as he (or she) paid cents for one of them. Each wife spent 75 cents more than her husband. Ann bought one more present that Bill Brown, Betty one less than Joe Jones. What is Mary's last name?

6.17. It was a very hot day and the 4 couples drank together 44 bottles of coca-cola. Ann had 2, Betty 3, Carol 4, and Dorothy 5 bottles. Mr. Brown drank just as many bottles as his wife, but each of the other men drank more than his wife: Mr. Green twice, Mr. White three times, and Mr. Smith four times as many bottles. Tell the last names of the four ladies.

6.18. *More problems.* Try to consider further examples from the viewpoint of this chapter. Pay attention to the division of the condition into clauses and weigh the advantages and the disadvantages of beginning the work with this or that clause. Review a few problems you have solved in the past with this viewpoint in mind and seek new problems in solving which this viewpoint has a chance to be useful.

6.19. *An intermediate goal.* We have started already working at our problem, but we are still in an initial phase of our work. We have understood our problem as a whole; it is a problem to find.   We have answered the question "What is the unknown?"; we know what kind of thing we are looking for. We have also listed the data and have understood the condition as a whole, and now we want to *split the condition into appropriate parts.*

Observe that this task need not be trivial: there may be several possibilities to subdivide the condition and we want, of course, the most advantageous subdivision.   For instance, in solving a geometric problem by algebra, we express each clause of the condition by an equation; different subdivisions of the condition into clauses yield different systems of equations and, of course, we want to pick the system that is most convenient to handle (cf. sect. 2.5(3) and 2.5(4)).

In the statement of the proposed problem, the condition may appear as an undivided whole or it may be divided into several clauses.   In either case we are facing a task: to split the condition into appropriate clauses in the first case and, in the second case, to split the condition into more appropriate clauses.   The subdivision of the condition may bring us nearer to the solution: it is an *intermediate goal,* very important in some cases.

6.20. *Graphical representation.*   We have expressed a relation, required by the condition of the problem, which involves specified unknowns by a symbolic equation [introduced in sect. 6.1(3)].   We may express such relations also graphically, by a diagram, and the graphical representation may contribute to a clearer conception of a system of such relations.

We represent an unknown by a small circle, a relation between unknowns by a small square, and we express the fact that a certain relation involves a certain unknown by joining the square representing the relation to the circle representing the unknown.   Thus diagram (*a*) in Fig. 6.3 represents a system of four relations between four unknowns; we see from it, for instance, that there is just one unknown involved in all four relations and just one relation involving all four unknowns; in fact, the diagram (*a*) and the system of four equations in sect. 6.1(1) express exactly the same state of affairs, the one in geometric language and the other in the language of formulas.   The crossing of lines in a point which lies outside the little circles and squares [as it happens once in diagram (*a*)] is immaterial; we can imagine, in fact, that only the little circles and squares lie in the plane of the paper and the connecting lines are drawn through space and have no point in common, although their projections on the plane of the paper may cross accidentally.

As the diagram (*a*), also the diagrams (*b*), (*c*), (*d*), and (*e*) of Fig. 6.3 represent systems of relations considered before: point out the section or example where they have been considered.

(Fig. 6.4 exemplifies another kind of diagrammatic representation which is "dual" to the foregoing: both relations and unknowns are represented by lines, a relation by a horizontal, an unknown by a vertical, line; iff a relation contains an unknown, the lines have a common point.   The same fact is expressed by (*c*) in Fig. 6.3 and Fig. 6.4, and the same holds for (*d*).

†An algebraic representation is suggested by Fig. 6.4: a matrix in which

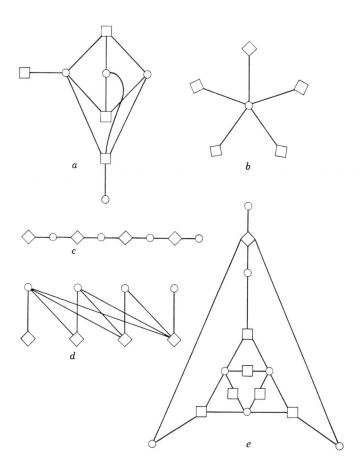

Fig. 6.3. Circles and squares, unknowns and relations.

each row corresponds to a relation and each column to an unknown of the system; an element of the matrix is 1 or 0 according as the relation concerned does or does not involve the unknown concerned.)

6.21. *Some types of nonmathematical problems.*   Which clause of the condition should we try to satisfy first? This question arises typically in various situations.   Having chosen a clause which appears to be of major importance, and having listed the objects (or some objects) satisfying this "major" clause, we bring into play the remaining "minor" clauses which remove most of the objects from the list and leave eventually one that satisfies also the minor clauses and so the full condition.   This pattern of procedure which we had opportunity to observe in the foregoing [sect. 6.3(3), ex. 6.15, ex. 6.16] is

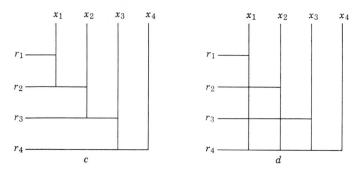

Fig. 6.4. Unknowns and relations, vertical and horizontal lines.

suitable for, and arises naturally in, various types of nonmathematical problems.

The translator's problem. In translating a French text into English, we have to find the correct English equivalent of a French word, for instance of the word "confiance." A French–English dictionary yields a list of English words (confidence, trust, reliance, assurance) which satisfy only a first, rather rough, clause of the full condition of our problem: we have to look carefully into the context to discover further, more subtle clauses hidden in it and bring these clauses into play to remove the less fitting words and choose the most appropriate one from the list.

Checkmate in two moves. There is given an arrangement of white and black chessmen on the chessboard, consistent with the rules of the game. The unknown is a move of white. The condition requires this move of white to be such that whatever move of black may follow there is a subsequent move of white that checkmates the black king.

The desired move of white has to "ward off" each possible move of black (prevent it from happening or prepare for answering it with a checkmate). And so the condition has as many clauses, we may say, as there are possible moves of black to be warded off.

A workable strategy is to begin with a crucial move of black which appears to involve a major threat and to list the moves of white capable of warding off this major threat. Then we consider other "minor" moves of black and remove from the list such moves of white which are not able to ward off one or the other "minor" move of black; the true solution should eventually remain alone on the list.

Engineering design. An engineer wants to design a new gadget. To be put into production, the new gadget has to fulfill a host of requirements; some are "technical" requirements such as smooth working, no danger for the user, durability, and so on, others are "commercial" requirements such as low price of manufacturing, sales appeal, and so on. The engineer retains at first the technical requirements (or some of them) which we may regard as constituting the "major" condition so that he has a clear-cut technical (physical) problem to solve. This problem has usually several solutions which the engineer lists and examines. This being done, the commercial requirements (which we have heretofore regarded as "minor") enter into play; they may discard many a smoothly operating gadget and may leave one the production of which appears the most profitable.

6.22. Without using paper and pencil, just by looking at it, solve the following system of three equations with three unknowns:

$$3x + y + 2z = 30$$
$$2x + 3y + z = 30$$
$$x + 2y + 3z = 30$$

Prove your solution.

6.23. Given a, b, and c, the lengths of the three sides of a triangle. Each vertex of the triangle is the center of a circle; these three circles are exterior to each other and touch each other. Find the three radii x, y, and z.

6.24. Find x, y, u, and v, satisfying the system of four equations:

$$y + u + v = -5$$
$$x + u + v = 0$$
$$x + y + v = -8$$
$$x + y + u = 4$$

Don't you see a short cut?

6.25. *A more refined classification.* The foregoing examples 6.22, 6.23, and 6.24 illustrate an important point: The circumstance that the condition of a problem involving several unknowns is *symmetric* with respect to these unknowns may, if recognized, influence the course of, and greatly facilitate, the solution. (Cf. also ex. 2.8 and MPR, vol. 1, pp. 187–188, ex. 41. Sometimes, as in ex. 6.23, we should consider not only permutations of the unknowns, but permutations of the unknowns *and* the data.) There are other cases, of less common occurrence but nevertheless interesting, in which the condition remains unchanged, not by all, but only by some permutations (by a certain group of permutations) of the unknowns (and the data). By following up this remark systematically, we would arrive at a still more refined classification of problems to find than the one implied by the basic remark of the present chapter [see sect. 6.1(6)] and we can foresee that such a classification would be of some interest for our study.

SOLUTIONS

Chapter 1

1.1. Circle with the given point as center and the given distance as radius.

1.2. Two straight lines parallel to the given line.

1.3. Straight line, the perpendicular bisector of the segment the endpoints of which are given.

1.4. Straight line parallel to the given parallels midway between them, that is, bisecting their distance.

1.5. Two straight lines, perpendicular to each other, bisectors of the angles included by the given lines.

1.6. Two circular arcs, symmetric to each other with respect to the line AB; they have the same endpoints, A and B.

1.8. Two loci, ex. 1.1.

1.9. Two loci, exs. 1.1, 1.2.

1.10. Two loci, exs. 1.2, 1.6.

1.11. Two loci, exs. 1.1, 1.6.

1.12. Two loci, ex. 1.5.

1.13. Two loci, ex. 1.2.

1.14. Two loci, exs. 1.1, 1.2.

1.15. Two loci, ex. 1.6.

1.16. By symmetry reduces to sect. 1.3(2), or to ex. 1.12.

1.17. Two loci, ex. 1.6.

1.18. (*a*) If X varies so that the areas of the two triangles $\triangle XCA$ and $\triangle XCB$ remain equal, the locus of X is the median passing through C. (Prove it!) The required point is the intersection of the medians. (*b*) If X varies so that the area of $\triangle ABX$ remains one third of the area of the given $\triangle ABC$, the locus

154

of X is a parallel to the side AB, at a distance equal to one third of the altitude dropped from C; see ex. 1.2. The required point is the intersection of such parallels to the sides. Both solutions use "two loci."

1.19. Join the center of the inscribed circle to both endpoints of the side a; in the triangle so obtained the angle at the center of the inscribed circle is

$$180° - \frac{\beta + \gamma}{2} = 90° + \frac{\alpha}{2}$$

Two loci, exs. 1.2, 1.6.

1.20. Auxiliary figure : the right triangle with hypotenuse a and leg h_b.

1.21. Auxiliary figure, see ex. 1.20.

1.22. Auxiliary figures, see ex. 1.20.

1.23. Auxiliary figure: right triangle with leg h_a and opposite angle β.

1.24. Auxiliary figures, see ex. 1.23. Other solution: see ex. 1.34.

1.25. Auxiliary figure: right triangle with hypotenuse d_α and height h_a.

1.26. Auxiliary figure: triangle from three sides.

1.27. Assume that a is longer than c. Auxiliary figure: triangle from sides $a - c, b, d$; see HSI, Variation of the problem 5, pp. 211–213.

1.28. Generalization of ex. 1.27 which corresponds to the case $\epsilon = 0$. Auxiliary figure: triangle from a, c, ϵ; see MPR, vol. 2, pp. 142–145.

1.29. Auxiliary figure: triangle from $a, b + c, \alpha/2$.

1.30. Auxiliary figure: triangle from $a, b + c, 90° + (\beta - \gamma)/2$.

1.31. Auxiliary figure: triangle from $a + b + c, h_a, \alpha/2 + 90°$. See HSI Auxiliary elements 3, pp. 48–50, and Symmetry, pp. 199–200.

1.32. Appropriate modification of the approach in sect. 1.6(1): let one of the two radii shrink at the same rate as the other expands. Auxiliary figure: tangents to a circle from an outside point, followed by the construction of two rectangles.

1.33. Cf. sect. 1.6(1). Auxiliary figure: circle circumscribed about the triangle the vertices of which are the centers of the three given circles.

1.34. Similar triangle from α, β; obtain afterwards the required size by using the given length d_y. Essentially the same for ex. 1.24.

1.35. Similar figures: center of similarity is the vertex of the right angle in the given triangle. The bisector of this right angle intersects the hypotenuse in a vertex of the desired square.

1.36. Generalization of ex. 1.35. Center of similarity is A (or B). Cf. HSI, section 18, pp. 23–25.

1.37. Similar figures: center of similarity is the center of the circle. The required square is symmetric with respect to the same line as the given sector.

1.38. Similar figure: any circle touching the given line, the center of which is on the perpendicular bisector of the segment joining the two given points. The point of intersection of this bisector and of the given line is center of similarity. Two solutions.

1.39. Symmetry with respect to the bisector of the appropriate angle included by the given tangents yields one more point through which the circle must pass, and so reduces the problem to ex. 1.38.

1.40. The radii of the inscribed circle drawn to the points of tangency include the angles $180° - \alpha$, $180° - \beta, \ldots$; hence a similar figure is immediately obtainable. Applicable to circumscribable polygons with any number of sides.

1.41. Let A denote the area and a, b, c the sides of the required triangle (ex. 1.7) so that

$$2A = ah_a = bh_b = ch_c$$

Construct a triangle from the given sides h_a, h_b, h_c and let A' denote its area and a', b', c' its corresponding altitudes so that

$$2A' = h_a a' = h_b b' = h_c c'$$

Therefore,

$$\frac{a}{a'} = \frac{b}{b'} = \frac{c}{c'}$$

and so the triangle with the easily obtainable sides a', b', c' is *similar* to the required triangle.

1.42. The foregoing solution of ex. 1.41 is imperfect: if $h_a = 156$, $h_b = 65$, $h_c = 60$, the required triangle does exist, but the auxiliary triangle with the given sides h_a, h_b, h_c does not.

One possible remedy is a *generalization*: let k, l, m be any three positive integers, and (the notation is *not* the same as in ex. 1.41) a', b', c' the altitudes in the triangle with sides kh_a, lh_b, mh_c; then

$$\frac{a}{ka'} = \frac{b}{lb'} = \frac{c}{mc'}$$

For example, a triangle with sides 156, 65, and $120 = 2 \times 60$ does exist.

1.43. From the center of the circumscribed circle, draw a line to one of the endpoints of the side a and a perpendicular to this side. You so obtain a right triangle with hypotenuse R, angle α, and opposite leg $a/2$. This yields a relation between a, α, and R: you can construct any one of the three if the two others are given. (The relation can also be expressed by the trigonometric equation $a = 2R \sin \alpha$.) If the data of the proposed problem do not satisfy this relation, the problem is impossible; if they do satisfy it, the problem is indeterminate.

1.44. (*a*) Triangle from α, β, γ: the problem is either impossible or indeterminate. (*b*) The general situation behind ex. 1.43 and (*a*): the existence of the solution implies a relation between the data and, therefore, the solution is either indeterminate or nonexistent according as the relation is, or is not, satisfied by the data. (*c*) By the solution of ex. 1.43 reduce to: triangle from a, β, α. (*d*) By the solution of ex. 1.43 reduce to ex. 1.19.

1.45. We disregard disturbances influencing the velocity of sound which we cannot control (as wind and varying temperatures). Then, from the time difference of the observations at the listening posts A and B, we obtain the difference of two distances, $AX - BX$ which yields a locus for X: a hyperbola. We obtain another hyperbola comparing C with A (or B) and the intersection of the two hyperbolas yields X.   Main analogy with ex. 1.15: the observations yield two loci.   Main difference: the loci are circular arcs there, but hyperbolas here.   We cannot describe a hyperbola with ruler and compasses, but we can describe it with some other gadget and a machine could be constructed to evaluate conveniently the observations of the three listening posts.

1.46. Those loci would not be usable if we took the statement of the pattern in sect. 1.2 literally.   In fact, those loci are useful and have been used several times in the foregoing examples, and it is the statement in sect. 1.2 that needs extension: we should admit a locus when it is a union of a finite number of straight lines, or circles, or segments of straight lines, or arcs of circles.

1.48. If the parts into which the condition is split are jointly equivalent to the condition, the various manners of splitting must be equivalent to each other.   Hence a theorem for the triangle: the perpendicular bisectors of the sides (there are three) pass through the same point.   And for the tetrahedron: the perpendicular bisecting planes of the edges (there are six) pass through the same point.

1.50. (1) Avoiding certain exceptional cases (see exs. 1.43 and 1.44) take any three different constituent parts of a triangle listed in ex. 1.7 as data and propose to construct a triangle.   Here are a few more combinations with which the construction is easy:

$$
\begin{array}{lll}
a, & h_b, & R \\
a, & h_b, & m_b \\
a, & h_b, & m_a \\
h_a, & d_\alpha, & b \\
h_a, & m_a, & m_b \\
h_a, & m_b, & m_c \\
h_a, & h_b, & m_a \\
a, & b, & R
\end{array}
$$

Also α, β, and any line not yet mentioned in ex. 1.24 or ex. 1.34.   Less easy

$$
\alpha, \quad r, \quad R
$$

(2) There are several problems about trihedral angles, similar to that discussed in sect. 1.6(3), which are important and can be solved without invoking explicitly the help of descriptive geometry.   Here is one: "Being given a, a face angle, and β and γ, the adjacent dihedral angles of a trihedral angle, construct b and c, the remaining face angles."   The solution is not difficult, but it would take up too much space to explain it here.

(3) Ex. 1.47 is the space analogue of sect. 1.3(1).   Discuss the space analogues of sect. 1.3(2), ex. 1.18, sect. 1.3(3), ex. 1.14.

No solution: **1.7, 1.47, 1.49, 1.51.**

Chapter 2

2.1. If Bob has x nickels and y dimes, we can translate the condition into the system of two equations

$$5x + 10y = 350$$
$$x + y = 50$$

which, after an obvious simplification, precisely coincides with the system in sect. 2.2(3).

2.2. There are m pipes to fill, and n pipes to empty, a tank. The first pipe can fill the tank in a_1 minutes, the second pipe in a_2 minutes, ... the pipe number m in a_m minutes. Of the other kind of pipes, the first can empty the tank in b_1 minutes, the second in b_2 minutes, ... the pipe number n in b_n minutes. With all pipes open, how long will it take to fill the empty tank?

The required time t satisfies the equation

$$\frac{t}{a_1} + \frac{t}{a_2} + \cdots + \frac{t}{a_m} - \frac{t}{b_1} - \frac{t}{b_2} - \cdots - \frac{t}{b_n} = 1$$

(If the solution t turns out negative, how do you interpret it? Possibly, there is no solution. How do you interpret this case?)

2.3. (a) Mr. Vokach (his name means "Smith" in Poldavian) spends one-third of his income on food, one-fourth on housing, one-sixth on clothing, and has no other expenses (there is no income tax in lucky Poldavia). He wonders how long he could live on one year's pay.

(b) What voltage should be maintained between two points connected by three parallel wires, the resistance of which is 3, 4, and 6 ohms, respectively, in order that the total current carried jointly by the three wires should be of intensity 1?

And so on.

2.4. (a) x remains unchanged if we substitute $-w$ for w: starting with the wind and returning against it the plane attains the same extreme point in a given time.

(b) Test by dimension; see HSI, pp. 202–205.

2.5. The system

$$x + y = v$$
$$ax + by = cv$$

agrees fully with that obtained in sect. 2.6(2).

2.6. Choose the coordinate system in the same relative position to the line AB as in sect. 2.5(1) and set $AB = a$. The required center (x, y) of the circle touching the four given arcs satisfies the two equations

$$a - \sqrt{x^2 + y^2} = \sqrt{\left(x - \frac{a}{4}\right)^2 + y^2} - \frac{a}{4}$$

$$x = \frac{a}{2}$$

from which follows

$$y = a\sqrt{6}/5$$

2.7. Heron's formula appears rather formidable, but is, in fact, quite manageable if you observe the combination "sum times difference" often enough:

$$
\begin{aligned}
16D^2 &= (a + b + c)(-a + b + c)(a - b + c)(a + b - c) \\
&= [(b + c)^2 - a^2][a^2 - (b - c)^2] \\
&= (2bc - a^2 + b^2 + c^2)(2bc + a^2 - b^2 - c^2) \\
&= 4b^2c^2 - (b^2 + c^2 - a^2)^2 \\
&= 4(p^2 + q^2)(p^2 + r^2) - (2p^2)^2
\end{aligned}
$$

2.8. (*a*) *Relevant knowledge.* Approach (3) supposes more knowledge of plane geometry (Heron's formula is less familiar than the expression of the area in terms of base and height). Yet approach (4) needs more knowledge of solid geometry (we have to see, and then to prove, that k is perpendicular to a).

(*b*) *Symmetry.* The three data A, B, and C play the same role, the problem is symmetric in A, B, and C. Approach (3) respects this symmetry, but approach (4) breaks with it and prefers A to B and C.

(*c*) *Planning.* Approach (3) proceeds more "methodically," we can follow it with some confidence from the start. And, in fact, it leads quite clearly to that system of seven equations which appeared to us, at the first blush, too formidable. [This is not the fault of the approach which hints, in fact, a procedure to solve them; see sect. 6.4(2).] It is less visible in advance that approach (4) will be helpful, but it "muddles through" somehow (thanks to a lucky remark) and attains the final result with a much shorter computation.

2.9. $V^2 = p^2q^2r^2/36 = 2ABC/9.$

2.10. From the three equations in sect. 2.5(3) that express a^2, b^2, and c^2 in terms of p, q, and r, we obtain

$$p^2 + q^2 + r^2 = S^2$$
$$p^2 = S^2 - a^2, \qquad q^2 = S^2 - b^2, \qquad r^2 = S^2 - c^2$$

and so ex. 2.9 yields

$$V^2 = (S^2 - a^2)(S^2 - b^2)(S^2 - c^2)/36$$

2.11. $d^2 = p^2 + q^2 + r^2$. This problem is broadly treated in HSI, Part I; see pp. 7–8, 10–12, 13–14, 16–19.

2.12. The notation chosen agrees both with ex. 2.11 and with sect. 2.5(3)—pay attention to both diagonals of the same face. Repeating a computation already done in ex. 2.10, we find

$$d^2 = (a^2 + b^2 + c^2)/2$$

2.13. A tetrahedron is determined by the lengths of its six edges—this results from the space analogue of the very first problem we have discussed in sect. 1.1. Yet we obtain the required configuration of the six edges, and so the proposed tetrahedron, in choosing one appropriate diagonal in each face of the box considered in ex. 2.11 and 2.12.   The volume of this box is pqr.   Cut off from the box four congruent tetrahedra, each with a trirectangular vertex and with volume $pqr/6$, see ex. 2.9; you obtain so the proposed tetrahedron whose volume is, therefore,

$$V = pqr - 4pqr/6 = pqr/3$$

Now, see ex. 2.10, $p^2 = S^2 - a^2$ and so on; hence

$$V^2 = (S^2 - a^2)(S^2 - b^2)(S^2 - c^2)/9$$

2.14. Ex. 2.10: If $V = 0$, one of the factors, for instance $S^2 - a^2 = p^2$ vanishes, and so two faces degenerate into line segments; the two other faces become coincident right triangles.

Ex. 2.13: If $V = 0$, the tetrahedron degenerates into a (doubly covered) rectangle; all four faces become congruent right triangles; in fact, $S^2 - a^2 = 0$ involves $a^2 = b^2 + c^2$.

2.15. As the last equation sect. 2.7 shows, the side x of the desired rectangle is the hypotenuse of a right triangle with legs $3a$ and a.   This segment x can be fitted into the cross in four different (but not essentially different) ways; its midpoint must coincide with the center of the cross, which divides it into two parts, each of the same length $x/2$ as the other side of the desired rectangle. All this suggests strongly the solution exhibited by Fig. S2.15.

2.16. (a) $x^2 = 12 \cdot 9 - 8 \cdot 1$, $x = 10$.

(b) Shift two units to the left and one unit upward, since

$$10 = 12 - 2 = 9 + 1$$

(c) Retention of the central symmetry is more likely.

All this suggests Fig. S2.16.

2.17. Let x and y be the loads carried by the mule and the ass, respectively. Then

$$y + 1 = 2(x - 1), \quad x + 1 = 3(y - 1); \qquad x = 13/5, \quad y = 11/5$$

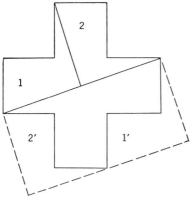

Fig. S2.15.

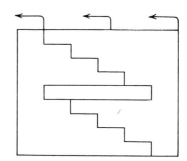

Fig. S2.16.

2.18. Mr. has h pounds, Mrs. w pounds, x pounds are free:
$$h + w = 94, \quad \frac{h - x}{1.5} = \frac{w - x}{2} = \frac{94 - x}{13.5}; \quad x = 40$$

2.19. 700, 500, $x = 400$ from
$$x + (x + 100) + (x + 300) = 1600$$

2.20. Each son receives 3000.

2.21. If the share of each child is x, and the whole fortune y, the shares of the children are

first: $\qquad x = 100 + \dfrac{y - 100}{10}$

second: $\qquad x = 200 + \dfrac{y - x - 200}{10}$

third: $\qquad x = 300 + \dfrac{y - 2x - 300}{10}$

and so on.

The difference of any two consecutive right hand sides is
$$100 - \frac{x + 100}{10}$$

If this difference equals 0 (as it should) $x = 900$, then (from the first equation) $y = 8100$: there were 9 children.

2.22. Let the three players initially have the amounts x, y, and z, respectively; it will be advantageous to consider
$$x + y + z = s$$
($s = 72$). We have to consider the amounts owned by the players at four different instants; any two consecutive instants are separated by a game; the total sum owned by the three is always s:

First	Second	Third
x	y	z
$2x - s$	$2y$	$2z$
$4x - 2s$	$4y - s$	$4z$
$8x - 4s = 24$	$8y - 2s = 24$	$8z - s = 24$

Hence $x = 39$, $y = 21$, $z = 12$.

2.23. Analogous to sect. 2.4(1) and 2.4(2), particular case of ex. 2.2 with

$$m = 3, \quad n = 0, \quad a_1 = 3, \quad a_2 = 8/3, \quad a_3 = 12/5$$

Hence $t = \frac{8}{9}$ of a week.

2.24. Newton means a generalization tending in the direction of ex. 2.2, but going less far: without "emptying pipes," there are no b's, $n = 0$.

2.25. Wheat, barley, and oats cost 5, 3, and 2 shillings a bushel, respectively. See ex. 2.26.

2.26. Let

$$x, \quad y, \quad z$$

be the prices of three commodities, and let p_ν be the price of the mixture in which

$$a_\nu, \quad b_\nu, \quad c_\nu$$

units of these commodities are contained, respectively, $\nu = 1, 2, 3$. We have thus a system of three equations

$$a_\nu x + b_\nu y + c_\nu x = p_\nu$$

$\nu = 1, 2, 3$. We obtain this generalization from the foregoing ex. 2.25 in passing from the array of numbers

40	24	20	312
26	30	50	320
24	120	100	680

to the array of letters

a_1	b_1	c_1	p_1
a_2	b_2	c_2	p_2
a_3	b_3	c_3	p_3

There is no difficulty in passing from 3 to n different commodities.

2.27. Let

α denote the quantity of grass per acre when the pasture starts to be used,
β the quantity of grass eaten by one ox in one week,
γ the quantity of grass that grows on one acre in one week,
a_1, a_2, a the number of oxen,
m_1, m_2, m the numbers of acres,
t_1, t_2, t the numbers of weeks in the three cases considered, respectively.
a, α, β, and γ are unknown, the remaining eight quantities are numerically given.

The conditions are

$$m_1(\alpha + t_1\gamma) = a_1 t_1 \beta$$
$$m_2(\alpha + t_2\gamma) = a_2 t_2 \beta$$
$$m(\alpha + t\gamma) = at\beta$$

a system of 3 equations for the 3 unknowns α/β, γ/β, a, which yields

$$a = \frac{m[m_1 a_2 t_2(t - t_1) - m_2 a_1 t_1(t - t_2)]}{m_1 m_2 t(t_2 - t_1)}$$

and, with the numerical data, $a = 36$.

2.28. Of an arithmetic progression
with five terms $a, a + d, \ldots, a + 4d$
find the first term a
and the difference d
being given that
the sum of all terms equals 100 $a + (a + d) + \cdots$
$$+ (a + 4d) = 100$$
and the sum of the last three terms $(a + 2d) + (a + 3d) + (a + 4d)$
equals 7 times the sum of the
first two terms $= 7[a + (a + d)]$

From the equations

$$5a + 10d = 100, \qquad 11a - 2d = 0$$

$a = 5/3$, $d = 55/6$ and so the progression is

$$10/6, \quad 65/6, \quad 120/6, \quad 175/6, \quad 230/6$$

2.29. $$\frac{m}{r} + m + mr = 19$$

$$\frac{m^2}{r^2} + m^2 + m^2 r^2 = 133$$

Set $$r + \frac{1}{r} = x$$

This changes the system into

$$m(x + 1) = 19, \qquad m^2(x^2 - 1) = 133$$

Division yields two *linear* equations for mx and m. Hence $m = 6$, $x = 13/6$, $r = 3/2$ or $2/3$; there are two (only trivially different) progressions: 4, 6, 9 and 9, 6, 4.

2.30. $$a(q^3 + q^{-3}) = 13, \quad a(q + q^{-1}) = 4$$

Division yields a quadratic for q^2. The progression is

$$1/5, \quad 4/5, \quad 16/5, \quad 64/5$$

or the same terms in reverse order.

2.31. Let x be the number of the partners. Express the profit of the partnership in two different ways (as received and as distributed):

$$(8240 + 40x \cdot x)\frac{x}{100} = 10x \cdot x + 224$$

The equation

$$x^3 - 25x^2 + 206x - 560 = 0$$

has no negative roots (substitute $x = -p$). If there is a rational root, it must be a positive integer, a divisor of 560. This leads to trying successively $x = 1, 2, 4, 5, 7, 8, 10, 14, 16, \ldots$ In fact, the roots are 7, 8, and 10. (Of course, Euler first made up the equation, then the story—you could try to imitate him.)

2.32. The centers of the four circles nonconcentric with the given square are vertices of another square of which we express the diagonal in two different ways:

$$(4r)^2 = 2(a - 2r)^2$$

and so

$$r = (\sqrt{2} - 1)a/2$$

2.33. Let $x + (d/2)$ stand for the height of the isosceles triangle perpendicular to the base. Then

$$\left(x + \frac{d}{2}\right)^2 + \left(\frac{b}{2}\right)^2 = s^2, \qquad x^2 + \left(\frac{b}{2}\right)^2 = \left(\frac{d}{2}\right)^2$$

Elimination of x yields

$$4s^4 - 4d^2s^2 + b^2d^2 = 0$$

2.34. (a) The equation is of the first degree in d^2 as well as in b^2, but of the second degree in s^2: Hence, the problem to find s may be reasonably regarded as more difficult than the other two.

(b) d has a positive value iff $4s^2 > b^2$.

b has a positive value iff $d^2 > s^2$.

s has two different positive values iff $d^2 > b^2$.

The reader can learn here several things. Newton comments on the solution of ex. 2.33 as follows: "And hence it is that Analysts order us to make no Difference between the given and sought Quantities. For since the same Computation agrees to any Case of the given and sought Quantities, it is convenient that they should be conceived and compared without any Difference... or rather it is convenient that you should imagine, that the Question is proposed of those *Data* and *Quaesita*, given and sought Quantities, by which you think it is most easy to make out your Equation." He adds a little later: "Hence, I believe, it will be manifest what Geometricians mean, when they bid you imagine that to be already done which is sought."

("Take the problem as solved"; cf. sect. 1.4.)

2.35. In setting up our equations we proceed in the direction just opposite to the one suggested by the surveyor's situation: We regard x and the angles $\alpha, \beta, \gamma, \delta$ as given, and l as the unknown. From $\triangle UVG$ we find GV in terms of x, $\alpha + \beta$, and γ (law of sines). From $\triangle VUH$ we find HV in terms of x, β, and $\gamma + \delta$ (law of sines). From $\triangle GHV$ we find l in terms of GV, HV, and δ (law of cosines) and, by using the expressions for GV and HV, we obtain

$$l^2 = x^2 \left[\frac{\sin^2(\alpha + \beta)}{\sin^2(\alpha + \beta + \gamma)} + \frac{\sin^2 \beta}{\sin^2(\beta + \gamma + \delta)} - \frac{2 \sin(\alpha + \beta) \sin \beta \cos \delta}{\sin(\alpha + \beta + \delta) \sin(\beta + \gamma + \delta)} \right]$$

Hence, express x^2 in terms of l, α, β, γ, and δ.

2.36. Let

A,	$2s$,	a,	b,	c

stand for

area,   perimeter,   hypotenuse,   remaining sides,

respectively; A and s are given, a, b, and c unknown. To solve the system

$$a + b + c = 2s, \qquad bc = 2A, \qquad a^2 = b^2 + c^2$$

express $(b + c)^2$ in two different ways:

$$(2s - a)^2 = a^2 + 4A$$

$$a = s - \frac{A}{s}$$

2.37. Lengths of the sides of the triangle $2a$, u, v; $u + v = 2d$; the altitude perpendicular to side of length $2a$ is of length h.

Given a, h, d, find u, v.

Introduce x and y, orthogonal projections of the sides u and v on $2a$, respectively, and z, where

$$x - y = 2z$$

Also

$$x + y = 2a$$

$$u^2 = h^2 + x^2 \qquad\qquad v^2 = h^2 + y^2$$

Hence,

$$u^2 - v^2 = x^2 - y^2$$

or

$$2d(u - v) = 2a \cdot 2z$$

$$u = d + \frac{a}{d}z \qquad\qquad\qquad v = d - \frac{a}{d}z$$

$$x = a + z \qquad\qquad\qquad\qquad y = a - z$$

$$\left(d + \frac{a}{d}z\right)^2 = h^2 + (a + z)^2$$

$$z^2 = d^2\left(1 - \frac{h^2}{d^2 - a^2}\right)$$

2.38. If a and b are the lengths of two nonparallel sides, and c and d the lengths of the two diagonals, then

$$2(a^2 + b^2) = c^2 + d^2$$

In fact, the diagonals dissect the parallelogram into four triangles: Apply the law of cosines to two neighboring triangles.

2.39. $(2b - a)x^2 + (4a^2 - b^2)(2x - a) = 0$

If $a = 10$, $b = 12$, then $x = 16(-8 + 3\sqrt{11})/7$, very nearly $32/7$. Interpret the case $a = 2b$.

2.40. $a^2(3 + \sqrt{3})/2$

2.41. $1/3$, $2/9$, $2/9$, $2/9$. In fact, the sides of the larger triangle are divided by the vertices of the inscribed triangle in the proportion 2 to 1.

2.42. (Stanford 1957.) Consider the simplest case first, that of the equilateral triangle. Symmetry may lead us to suspect that in this case the four triangular pieces will also be equilateral. If this is so, however, the sides of the triangular pieces must be *parallel* to the sides of the given triangle: with this remark, we have discovered the essential feature of the configuration that solves the problem not only in the particular case examined, but also in the general case. (We pass from the equilateral triangle to the general triangle by "affinity.") By four parallels to a side of the given triangle, dissect each of the other two sides into five equal segments. Performing this construction three times, with respect to each side of the given triangle, we divide it into 25 congruent triangles similar to it. From these 25 triangular pieces, we easily pick out the four mentioned in the problem: the area of each of them is 1/25 of the given triangle's area. (This solution is uniquely determined; I omit the proof.)

2.43. (Stanford 1960.) Generalize: The point P lies in the interior of a rectangle, its distances from the four corners are a, b, c, and d, from the four sides x, y, x', y', in cyclical order (as they are met by the hands of a watch). Then, with appropriate notation

$$a^2 = y'^2 + x^2, \quad b^2 = x^2 + y^2, \quad c^2 = y^2 + x'^2, \quad d^2 = x'^2 + y'^2$$

and so

$$a^2 - b^2 + c^2 - d^2 = 0$$

In our case $a = 5$, $b = 10$, $c = 14$, and so

$$d^2 = 25 - 100 + 196 = 121, \quad d = 11$$

Observe that the data a, b, and c which determine d are insufficient to determine the sides $x + x'$ and $y + y'$ of the rectangle.

2.44. (I) Let s stand for the side of the square. Then, by ex. 2.43, $x + x' = y + y' = s$, and we have three equations for the three unknowns x, y, and s:

$$x^2 + (s - y)^2 = a^2, \quad x^2 + y^2 = b^2, \quad y^2 + (s - x)^2 = c^2$$

Hence,

$$2sy = s^2 + b^2 - a^2, \quad 2sx = s^2 + b^2 - c^2$$

and, by squaring and adding, we find

$$s^4 - (a^2 + c^2)s^2 + [(b^2 - a^2)^2 + (b^2 - c^2)^2]/2 = 0$$

a quadratic equation for s^2.

(II) Check the geometric meaning of the particular cases:

(1) $s^2 = 2a^2$ or $s = 0$.
(2) $s = a$.
(3) s imaginary unless $c^2 = 2b^2 = 2s^2$.
(4) s imaginary unless $a^2 = c^2 = s^2$.

2.45. (Stanford 1959.) $100\pi/4$ and $100\pi/(2\sqrt{3})$ or approximately 78.54% and 90.69%, respectively. The transition from a large (square) table to the infinite plane involves, in fact, the concept of limit on which, however, we do not insist as the result is intuitive.

2.46. In following the procedure of ex. 2.32, we express the diagonal of an appropriate cube in two different ways:

$$(4r)^2 = 3(a - 2r)^2$$
$$r = (2\sqrt{3} - 3)a/2$$

2.47. The four vertices of a rectangle, taken in cyclical order, are at the distance a, b, c, and d, respectively, from a point P (which may be located anywhere in space). Being given three of these distances, find the fourth one. The relation

$$a^2 - b^2 + c^2 - d^2 = 0$$

found in ex. 2.43 remains valid in the present more general situation, and the solution follows from it immediately. This can be applied, for instance, to a point P and four appropriately chosen vertices of a box (rectangular parallelepiped) since any two diagonals of a box are also the diagonals of a certain rectangle.

2.48. The solution of a problem of solid geometry often depends on a "key plane figure" which opens the door to the essential relations.

Through the altitude of the pyramid, pass a plane that is parallel to two sides of the base (and perpendicular to its two other sides). The intersection of this plane with the pyramid is an isosceles triangle which can be used as key figure: its height is h, its base, say a, is equal in length to a side of the base of the pyramid, its legs are of length $2a$, since each one is the height of a lateral face. Hence,

$$(2a)^2 = \left(\frac{a}{2}\right)^2 + h^2$$

and so the desired area is

$$5a^2 = 4h^2/3$$

2.49. For instance: The area of the surface of a regular pyramid equals four times the area of its hexagonal base. Given a, the length of a side of this base, find h, the height of the pyramid ($h = \sqrt{6}a$).

See also ex. 2.52.

2.50. In a parallelogram the sum of the squares of the 2 diagonals equals the sum of the squares of the 4 sides. (Restatement of the result of ex. 2.38.)

In a parallelepiped, let

$$D, \qquad\qquad E, \qquad\qquad F$$

stand for the sum of the squares of the

| 4 diagonals, | 12 edges, | 12 face-diagonals |

respectively. Then

$$D = E = F/2$$

(Follows from the result of ex. 2.38 by repeated application.)

2.51. The square of the desired area is

$$16s(s - a)(s - b)(s - c)$$

This may be regarded as an analogue to Heron's theorem, but is too close to it to be interesting.

2.52. (Stanford 1960.) Let a stand for the side of a triangle, T for the volume of the tetrahedron, and O for the volume of the octahedron.

First solution. The octahedron is divided by an appropriate plane into two congruent regular pyramids with a common square base the area of which is a^2. The height of one of these pyramids is $a/\sqrt{2}$ (the "key plane figure" passes through a diagonal of the base) and so

$$O = 2\frac{a^2}{3}\frac{a}{\sqrt{2}} = \frac{a^3\sqrt{2}}{3}$$

Pass a plane through the altitude (of length h) of the tetrahedron and through a coterminal edge; the intersection (the key plane figure) shows two right triangles from which

$$h^2 = a^2 - \left(\frac{2a\sqrt{3}}{6}\right)^2 = \left(\frac{a\sqrt{3}}{2}\right)^2 - \left(\frac{a\sqrt{3}}{6}\right)^2$$

$$= \frac{2a^2}{3}$$

and so

$$T = \frac{1}{3}\frac{a}{2}\frac{a\sqrt{3}}{2}\frac{a\sqrt{2}}{\sqrt{3}} = \frac{a^3\sqrt{2}}{12}$$

Finally

$$O = 4T$$

Second solution. Consider the regular tetrahedron with edge $2a$; its volume is 2^3T; four planes, each of which passes through the midpoints of three of its edges terminating in the same vertex, dissect it into four regular tetrahedra, each of volume T, and a regular octrahedron of volume O. Hence,

$$4T + O = 8T$$

which yields again $O = 4T$.

2.53. The volumes are as

$$\frac{1}{a} : \frac{1}{b} : \frac{1}{c}$$

the surface areas as

$$\frac{b + c}{a} : \frac{c + a}{b} : \frac{a + b}{c}$$

2.54. (Stanford 1951.) The difference of the volumes, frustum minus cylinder,

$$\pi h\left[\frac{a^2 + ab + b^2}{3} - \left(\frac{a + b}{2}\right)^2\right] = \frac{\pi h(a - b)^2}{12}$$

is positive unless $a = b$ and the solids coincide.

MPR, vol. I, chapter VIII, contains several applications of algebraic inequalities to geometry.

2.55. Let r be the radius of the circle circumscribed about $\triangle ABC$. Then

$$r^2 = h(2R - h), \qquad r = \frac{2}{3}\frac{\sqrt{3}a}{2}$$

and so

$$R = \frac{a^2}{6h} + \frac{h}{2}$$

The term $h/2$ is often negligible in practice.

2.57. 35 miles; see ex. 2.58.

2.58. We list each given numerical value in parenthesis following the letter that generalizes it:

a (7/2) is the velocity of A,

b (8/3) the velocity of B,

c (1) the number of hours that pass between the two starts,

d (59) the distance between the two starting points.   Then

$$x + y = d, \qquad \frac{x}{a} - \frac{y}{b} = c, \qquad x = \frac{a(bc + d)}{a + b}$$

Newton formulates the generalized problem as follows: "Having given the Velocities of two moveable Bodies, A and B, tending to the same Place, together with the Interval or Distance of the Places and Times from, and in which, they begin to move; to determine the Place they shall meet in."

2.59. (Stanford 1959.)   We use the notation

u for Al's speed,

v for Bill's speed,

t_1 for the time (counted from the start) the boys meet the first time,

t_2 for the time they meet the second time,

d the desired distance of the two houses.   Then

$$ut_1 = a, \qquad\qquad ut_2 = d + b$$
$$vt_1 = d - a, \qquad vt_2 = 2d - b$$

(1) By expressing u/v in two different ways, we obtain

$$\frac{a}{d - a} = \frac{d + b}{2d - b}$$

Hence, after discarding the vanishing root, we find $d = 3a - b$.

(2) Of course, Al.   Numerically: $u/v = 3/2$.

2.60. (Stanford 1955.) See ex. 2.61; see also HSI, pp. 236, 239–240, 247: problem 12.

2.61. Between the start and the first point where all $n + 1$ friends meet again, there are $2n - 1$ different phases:

        (1) Bob rides with A

        (2) Bob rides alone

        (3) Bob rides with B

        (4) Bob rides alone

.   .   .   .   .   .

$(2n - 1)$ Bob rides with L

Fig. S2.61, where $n = 3$, exhibits 5 phases; the lines representing the travels of A, B, or C are marked with the proper letters; the steeper slope renders the line

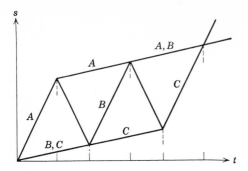

Fig. S2.61.

that represents the itinerary of the car easily recognizable. We see from the symmetry of the arrangement (especially clear from Fig. S2.61) that all n odd-numbered phases are of the same duration, say T, and all $n - 1$ even-numbered phases of the same duration, say T'. Express the total progress through the $2n - 1$ phases [in $nT + (n - 1)T'$ units of time] in two different ways (look first at Bob, then at one of his friends):

$$nTc - (n - 1)T'c = Tc + (n - 1)(T + T')p$$

whence

$$\frac{T}{T'} = \frac{c + p}{c - p}$$

(1) Rate of progress of the company is

$$\frac{nTc - (n - 1)T'c}{nT + (n - 1)T'} = c\,\frac{c + (2n - 1)p}{(2n - 1)c + p}$$

(2) The fraction of time when the car carries Bob alone

$$\frac{(n - 1)T'}{nT + (n - 1)T'} = \frac{(n - 1)(c - p)}{(2n - 1)c + p}$$

(3) Results (1) and (2) become intuitive in the extreme cases [just (2) for $n = \infty$ is less immediate]:

	$p = 0$	$p = c$	$n = 1$	$n = \infty$
(1) Rate of progress	$c/(2n - 1)$	c	c	p
(2) Fraction when Bob alone	$(n - 1)/(2n - 1)$	0	0	$(c - p)/2c$

2.62. Let t_1 be the time of descent of the stone and t_2 the time of ascent of the sound. From

$$T = t_1 + t_2, \qquad d = gt_1^2/2, \qquad d = ct_2$$

we find

$$d = \{-c(2g)^{-1/2} + [c^2(2g)^{-1} + cT]^{1/2}\}^2$$

Cf. MPR, vol. I, p. 165 and p. 264, ex. 29.

2.63. Introduce $\beta' = \angle ACO$. Since

$$\frac{\sin \omega}{\sin \beta} = \frac{AB}{AO}, \qquad \frac{\sin \omega'}{\sin \beta'} = \frac{AC}{AO}$$

$$\frac{\sin \omega}{\sin \omega'} \frac{\sin \beta'}{\sin \beta} = \frac{t}{t'}$$

On the other hand, $\beta' = \beta - (\omega' - \omega)$. Expressing $\sin \beta'/\sin \beta$ in two different ways, we obtain

$$\cot \beta = \cot(\omega' - \omega) - \frac{t \sin \omega'}{t' \sin \omega \sin (\omega' - \omega)}$$

2.64. Adding the three equations, we obtain

$$0 = a + b + c$$

If this relation is *not* satisfied by the data a, b, and c, "the problem is impossible," that is, there are no numbers x, y, z fulfilling the three simultaneous equations. If that relation *is* satisfied, the problem is *indeterminate*, that is, there are infinitely many solutions: from the first two equations

$$x = z + (3a + b)/7$$
$$y = z + (2a + 3b)/7$$

where z remains arbitrary.

Cf. ex. 1.43 and 1.44.

2.65. (Stanford 1955.)   Comparing coefficients of like powers on both sides of the identity, we obtain 5 equations

$$1 = p^2, \qquad 4 = 2pq, \qquad -2 = q^2 + 2pr, \qquad -12 = 2qr, \qquad 9 = r^2$$

for our 3 unknowns p, q and r. The first equation yields $p = \pm 1$, whence the following 2 equations, used successively, determine two systems of solutions

$$p = 1, \quad q = 2, \quad r = -3, \qquad \text{and} \qquad p = -1, \quad q = -2, \quad r = 3$$

both of which happen to satisfy also the remaining two equations.

Usually, it will not be possible to extract the square root, since usually it is impossible to satisfy a system with more equations than unknowns.

2.66. (Stanford 1954.)   Expanding the right-hand side of the hypothetical identity and equating corresponding coefficients, we obtain

(1) $aA = bB = cC = 1$
(2) $bC + cB = cA + aC = aB + bA = 0$

We derive from (2) that

$$bC = -cB, \qquad cA = -aC, \qquad aB = -bA$$

and multiplying these three equations, we derive further that

$$abcABC = -abcABC$$
$$abcABC = 0$$

Yet we derive from (1) that

$$abcABC = 1$$

This contradiction shows that the hypothetical identity from which we have started is impossible.

We have shown here that a system of 6 equations with the 6 unknowns a, b, c, A, B, and C is inconsistent.

2.67. $x = 5t, \qquad y = 60 - 18t, \qquad z = 40 + 13t$

are positive iff $0 < t < 60/18$. This leaves for t the values 1, 2, 3 and for (x, y, z) the three systems

$$(5, 42, 53), \qquad (10, 24, 66), \qquad (15, 6, 79)$$

2.68. Follow ex. 2.67: the system

$$x + y + z = 30$$
$$14x + 11y + 9z = 360$$

is satisfied by

$$x = 2t, \qquad y = 45 - 5t, \qquad z = 3t - 15$$
$$t = 5, 6, 7, 8, \text{ or } 9$$

2.69. $\qquad\qquad\qquad 100 + x = y^2, \qquad 168 + x = z^2$

Subtracting we obtain

$$(z - y)(z + y) = 68$$

Since $68 = 2^2 17$ can be decomposed into a product of two factors in just three ways:

$$68 = 1 \cdot 68 = 2 \cdot 34 = 4 \cdot 17$$

and y and z must be both odd or both even, there is just one solution:

$$z - y = 2, \qquad z + y = 34, \qquad z = 18, \qquad y = 16, \qquad x = 156$$

2.70. (Stanford 1957.) Bob has x stamps of which y sevenths are in the second book; x and y are positive integers,

$$\frac{2x}{10} + \frac{yx}{7} + 303 = x$$

and hence

$$x = \frac{3 \cdot 5 \cdot 7 \cdot 101}{28 - 5y}$$

The denominator on the right-hand side must be positive and *odd* since it must divide the numerator which is odd. This leaves three possibilities: $y = 1, 3$, and 5, and only the last one is suitable: $y = 5$ and $x = 3535$ are uniquely determined.

2.71. (Stanford 1960.) If the reduced price is x cents and there are y pens in the remaining stock, $x < 50$ and

$$xy = 3193$$

Now, $3193 = 31 \times 103$ is a product of two prime factors, and so it has precisely four different factors, 1, 31, 103, and 3193.   If we *assume* that x is an integer, $x = 1$, or 31.   If we *assume also* that $x > 1$, then $x = 31$.

2.75. (1) *Inconsistence*: Either there are, among the three planes, two which are different and parallel; or any two planes intersect and the three lines of intersection are different and parallel.

(2) *Dependence*: The plans possess a common straight line; two, or even all three, may coincide.

(3) *Consistence and independence*: There is just one common point, *the* point of intersection.

2.78. The current textbooks for secondary schools contain "word problems" in great number, although not in great variety.   Just such applications and such kinds of questions are usually lacking as could shed light on the general interest of the "Cartesian pattern."

From the foregoing examples, the reader can learn to attach useful questions to a problem that he has just solved.   I list a few such questions, referring to one illustrative example for each (the reader should look for further illustrations):

Can you check the result?   (Ex. 2.4.)

Check extreme (degenerate, limiting) cases.   (Ex. 2.14.)

Can you derive the result differently?   Compare the different approaches.   (Ex. 2.8.)

Could you devise another interpretation of the result?   (Ex. 2.3.)

Generalize the problem.   (Ex. 2.2.)

Devise an analogous problem.   (Ex. 2.47.)

Starting from any problem and asking the foregoing and similar questions, the reader may evolve new problems and find perhaps some interesting and not too difficult problem.   At any rate, in so asking, he has a good chance to deepen his understanding of the problem he started from and to improve his problem-solving ability.

Here are just two (not too easy) problems evolved from the foregoing.

(I) Check the result of ex. 2.35

(1) in supposing $\alpha = \delta$, $\beta = \gamma$, $\alpha + \beta = 90°$;

(2) in supposing $\alpha = \delta$, $\beta = \gamma$, but without prescribing a value for $\alpha + \beta$;

(3) by substituting δ, γ, β, and α for α, β, γ, and δ, respectively.

(II) Consider problems of solid geometry analogous to ex. 2.45.   (There is a hint in ex. 3.39.)

No solution: **2.56, 2.72, 2.73, 2.74, 2.76, 2.77.**

Chapter 3

3.1. For $n = 0$ and $n = 1$ the assertion is obvious.   Assume that it is valid for some value of n:

$$(1 + x)^n = 1 + \cdots + \binom{n}{r-1}x^{r-1} + \binom{n}{r}x^r + \cdots + x^n$$

Multiplying both sides by $1 + x$ you obtain

$$(1 + x)^{n+1} = 1 + \cdots + \left[\binom{n}{r} + \binom{n}{r-1} \right] x^r + \cdots + x^{n+1}$$

By virtue of the *recursion formula* of sect. 3.6(2), the coefficient of x^r in $(1 + x)^{n+1}$ turns out equal to

$$\binom{n+1}{r}$$

and so the binomial theorem, supposed to be valid for n, turns out to be valid also for $n + 1$. Observe that we have also used the *boundary condition* of sect. 3.6(2). Where?

3.2. Assume the result of ex. 3.1, set

$$x = \frac{b}{a}$$

and consider

$$a^n(1 + x)^n = (a + b)^n$$

3.3. Consider the assertion "S_p is a polynomial of degree $p + 1$" as a conjecture (what it originally was). This conjecture is certainly true in the first particular cases, $p = 0$, 1, and 2 (which suggested it; see the beginning of sect. 3.3). Let us *assume* that the conjecture is verified up to the case $p = k - 1$, that is, for $p = 0, 1, 2, \ldots, k - 1$ (for $S_0, S_1, S_2, \ldots, S_{k-1}$.) *Then* we can conclude (look at the final equation of sect. 3.4) that

$$\binom{k+1}{2} S_{k-1} + \binom{k+1}{3} S_{k-2} + \cdots + S_0 = P$$

(we introduce P as an abbreviation) is a polynomial of degree k. Now, we derive from that equation that

(1) $$S_k = \frac{(n+1)^{k+1} - 1 - P}{k+1}$$

Since P is of degree k in n, the highest term of $(n + 1)^{k+1}$ which is n^{k+1} cannot be canceled, and so our formula shows that S_k is a polynomial of degree $k + 1$ in n. We reached this conclusion assuming that S_0 is of degree 1, S_1 of degree 2,... and S_{k-1} of degree k.

To put it intuitively, the property considered of S_k (that it is of degree $k + 1$) has an "unquenchable tendency to propagate itself." We knew pretty early that S_0, S_1, and S_2 have this property; therefore, by our foregoing proof, also S_3 must have it; by the same proof also S_4 must have it, then S_5, and so on.

That the highest term in S_k is of the asserted form, is also obvious now from the formula (1).

Some of the following problems provide other approaches to the result just proved; see also ex. 4.2–4.7.

3.4. $$S_4 = S_2 \frac{6S_1 - 1}{5}$$

$$= \frac{n(n + 1)(2n + 1)(3n^2 + 3n - 1)}{30}$$

The proof by mathematical induction follows the standard pattern; see MPR, vol. 1, pp. 108–120.

3.5. Pattern suggested by sect. 3.2, 3.3, and 3.4, and by ex. 3.3.

3.6. Analogous to sect. 3.4:

$$n^k - (n - 1)^k = \binom{k}{1}n^{k-1} - \binom{k}{2}n^{k-2} + \cdots + (-1)^{k-1}\binom{k}{k}$$

3.7. Analogous to sect. 3.4:

$$[n(n + 1)]^k - [(n - 1)n]^k = n^k[(n + 1)^k - (n - 1)^k]$$

$$= 2\binom{k}{1}n^{2k-1} + 2\binom{k}{3}n^{2k-3} + 2\binom{k}{5}n^{2k-5} + \cdots$$

3.8. Analogous to sect. 3.4:

$$(2n + 1)[n(n + 1)]^k - (2n - 1)[(n - 1)n]^k$$
$$= n^k[(n + 1)^k + (n - 1)^k] + 2n^{k+1}[(n + 1)^k - (n - 1)^k]$$

$$= 2\left[\binom{k}{0} + 2\binom{k}{1}\right]n^{2k} + 2\left[\binom{k}{2} + 2\binom{k}{3}\right]n^{2k-2} + \cdots$$

3.9. From ex. 3.7 by recursion and mathematical induction.

3.10. From ex. 3.8 by recursion and mathematical induction.

3.11. By virtue of ex. 3.9 and 3.10, it is enough to verify the assertion for $S_1(x)$ and

$$S_2(x) = S_1(x)\frac{2x + 1}{3}$$

3.12. (*a*) By "little Gauss's method" (first approach of sect. 3.1):

$$[1 + (2n - 1)] + [3 + (2n - 3)] + \cdots = 2n\cdot\frac{n}{2} = n^2$$

(*b*) By the second approach of sect. 3.1; see the following.

(*c*) Generalize: consider the sum of an arithmetic series with initial term a, difference d, and n terms:

$$S = a + (a + d) + (a + 2d) + \cdots + (a + (n - 1)d)$$

Set the last term $a + (n - 1)d = b$; then (this is the second approach of sect. 3.1)

$$S = a + (a + d) + (a + 2d) + \cdots + (b - 2d) + (b - d) + b$$
$$S = b + (b - d) + (b - 2d) + \cdots + (a + 2d) + (a + d) + a$$

By adding and dividing by 2

$$S = \frac{a + b}{2}n$$

Specialize: $a = 1$, $b = 2n - 1$; then

$$S = \frac{1 + (2n - 1)}{2}n = n^2$$

(4) Look at Fig. 3.9.

(5) See ex. 3.13.

3.13. $1 + 4 + 9 + 16 + \cdots + (2n - 1)^2 + (2n)^2 - 4(1 + 4 + \cdots + n^2)$

$$= \frac{2n(2n + 1)(4n + 1)}{6} - 4 \cdot \frac{n(n + 1)(2n + 1)}{6}$$

$$= \frac{n(4n^2 - 1)}{3}$$

3.14. Follow the pattern of ex. 3.13:

$$\frac{4n^2(2n + 1)^2}{4} - 8 \frac{n^2(n + 1)^2}{4} = n^2(2n^2 - 1)$$

3.15. Use the notation of ex. 3.11:

$$1^k + 3^k + \cdots + (2n - 1)^k = S_k(2n) - 2^k S_k(n)$$

3.16. *More questions may be easier to answer than just one question.* (This is the "inventor's paradox"; see HSI, p. 121.)   Along with the proposed

$$2^2 + 5^2 + 8^2 + \cdots + (3n - 1)^2 = U$$

consider

$$1^2 + 4^2 + 7^2 + \cdots + (3n - 2)^2 = V$$

Then (suggested by ex. 3.15)

$$U + V + 9S_2(n) = S_2(3n)$$

Moreover

$$U - V = 3 + 9 + 15 + \cdots + (6n - 3) = 3n^2$$

We have here a system of two linear equations for the two unknowns U and V which yields not only the required

$$U = n(6n^2 + 3n - 1)/2$$

but also

$$V = n(6n^2 - 3n - 1)/2$$

For another method see ex. 3.17.

3.17. (See Pascal, l.c. footnote 3 in chapter 3.)   Generalizing the notation of sect. 3.3 (where we dealt with the particular case $a = d = 1$) we set

$$S_k = a^k + (a + d)^k + (a + 2d)^k + \cdots + [a + (n - 1)d]^k$$

Obviously, $S_0 = n$.   Substituting $1, 2, 3, \ldots, n$ for n in the relation

$(a + nd)^{k+1} - [a + (n - 1)d]^{k+1}$

$$= \binom{k + 1}{1}[a + (n - 1)d]^k d + \binom{k + 1}{2}[a + (n - 1)d]^{k-1}d^2 + \cdots$$

and adding, we obtain

$$(a + dn)^{k+1} - a^{k+1} = \binom{k + 1}{1}S_k d + \binom{k + 1}{2}S_{k-1}d^2 + \cdots + S_0 d^{k+1}$$

Hence we find $S_1, S_2, \ldots, S_k$ one after the other, recursively.   Do in detail the case $a = 2, d = 3, k = 2$; see ex. 3.16.

3.18. The sum required is

$$\frac{1 \cdot 2}{2} 2 + \frac{2 \cdot 3}{2} 3 + \frac{3 \cdot 4}{2} 4 + \cdots + \frac{(n-1)n}{2} n$$

$$= \tfrac{1}{2}[(2^3 - 2^2 + 3^3 - 3^2 + 4^3 - 4^2 + \cdots + n^3 - n^2)]$$

$$= \tfrac{1}{2}(S_3 - S_2) = \frac{(n-1)n(n+1)(3n+2)}{24}$$

by the results of sect. 3.2 and 3.3.

3.19. (a) $\dfrac{n(n^2 - 1)}{6}$;   (b) $1^{n-1}2^{n-2}3^{n-3}\ldots(n-1)^1$;   (c) $\dfrac{n^2(n^2 - 1)}{12}$

3.20. We have already computed E_1 in sect. 3.1 and E_2 in ex. 3.18.   A more efficient procedure is based on a classical fact of algebra: the elementary symmetric functions can be expressed in terms of the sums of like powers:

$$E_1 = S_1$$
$$E_2 = (S_1{}^2 - S_2)/2$$
$$E_3 = (S_1{}^3 + 2S_3 - 3S_1S_2)/6$$
$$E_4 = (S_1{}^4 + 3S_2{}^2 + 8S_1S_3 - 6S_1{}^2S_2 - 6S_4)/24$$

Combine these with our former results (sect. 3.1, 3.2, and 3.3, ex. 3.4).   Combining certain properties of the general expression of E_k in terms of $S_1, S_2, \ldots,$ S_k ("isobaric") with ex. 3.9 and 3.10, we can obtain not only the degree but also the coefficient of the highest term,

$$E_k(n) = \frac{n^{2k}}{k! 2^k} + \cdots$$

and we can derive that, for $k \geq 2$, $E_k(n)$ is divisible by

$$(n - k + 1)(n - k + 2)\ldots(n - 1)[n(n + 1)]^{[3 - (-1)^k]/2}$$

3.21. *Procedure (a) is a particular case of procedure (b).*   In fact, if A_{n+1} is implied by A_n alone, it is *a fortiori* (even with stronger reason) implied by $A_1, A_2, \ldots, A_{n-1}$ and A_n together.   That is, if statement (II*a*) happens to be correct, statement (II*b*) must be correct.   Hence, if we accept procedure (*b*), we are obliged to accept procedure (*a*).

Procedure (b) can be reduced to procedure (a).   Define B_n as the simultaneous assertion of the n propositions $A_1, A_2, \ldots, A_{n-1}$ and A_n.   Then

statement (I) means: B_1 *is true.*

statement (II*b*) boils down to: B_n *implies* B_{n+1}.

Hence, the statements (I) and (II*b*) about the sequence $A_1, A_2, A_3, \ldots$ boil down to statements (I) and (II*a*) with B_n substituted for A_n, (for $n = 1, 2, 3, \ldots$, of course).

3.22. Fig. 3.3 can be conceived as representing the case in which Bernie, Charlie, Dick, Roy, and Artie (blocks from northwest to southeast) put up the tent and the other five boys (Ricky, Abe, Al, Alex, and Bill—blocks from northeast to southwest) cook the supper.   Starting from this concrete case,

you should be able to see that to each division of the ten boys into two differently labeled teams of five there corresponds a shortest zigzag path from top
to bottom in Fig. 3.3 and, conversely, to each zigzag path of this kind there
corresponds such a division; the correspondence is one-to-one. Therefore,
the desired number of divisions is 252; see Fig. 3.3.

3.23. We are facing here a general situation, a *representative special case* of
which (MPR, vol. 1, p. 25, ex. 10) appears in ex. 3.22 and Fig. 3.3.

Number the individuals from 1 to n and let correspond the kth "base"
(horizontal row) of the Pascal triangle to the kth individual. An individual
belongs to the subset if, and only if, the zigzag path arrives at the corresponding base coming down along a block running *from northwest to southeast*.
In this manner, any subset of size r contained in the given set of size n can be
visualized as a zigzag path ending in a fixed point, and we count the subsets by
counting the zigzag paths. Cf. MPR, vol. 2, pp. 105–106, ex. 31.

3.24. $\dfrac{n(n-1)}{1\cdot 2}$ straight lines, $\dfrac{n(n-1)(n-2)}{1\cdot 2\cdot 3}$ triangles.

3.25. Given n points in space in "general position", there are

$$\frac{n(n-1)(n-2)(n-3)}{1\cdot 2\cdot 3\cdot 4}$$

tetrahedra with vertices chosen among the given n points.

3.26. $$\binom{n}{2} - n = \frac{n(n-3)}{2}$$

3.27. Two diagonals intersecting *inside* the given convex polygon are the
diagonals of a *convex* quadrilateral the four vertices of which are chosen among
the n vertices of the given polygon. Therefore the number of the intersections
in question is

$$\frac{n(n-1)(n-2)(n-3)}{1\cdot 2\cdot 3\cdot 4}$$

3.28. The red face can be chosen in

$$\binom{6}{1} = 6$$

different ways. From the remaining five faces, the two blue faces can be
chosen in

$$\binom{5}{2} = 10$$

different ways. Hence the total number of possibilities for distributing the
three colors among the six faces in the required manner is

$$\binom{6}{1}\binom{5}{2} = 6 \times 10 = 60$$

3.29. $\quad \dbinom{n}{r}\dbinom{s+t}{s} = \dfrac{n!}{r!(n-r)!}\cdot\dfrac{(s+t)!}{s!t!} = \dfrac{n!}{r!s!t!}$

3.30. A set of n individuals is divided into h nonoverlapping subsets (that is, two different subsets have no member in common); the first subset has r_1 members, the second r_2 members, ... and the last subset has r_h members so that

$$r_1 + r_2 + r_3 + \cdots + r_h = n$$

There are

$$\frac{n!}{r_1!r_2!r_3!\ldots r_h!}$$

different subdivisions of this kind. The numbering or labeling of the subsets is essential: if some of the numbers $r_1, r_2, \ldots, r_h$ happen to be equal, we must carefully distinguish between differently labeled subsets of equal size. Thus in ex. 3.22, we distinguish between the five individuals who put up the tent and the other five who cook the supper; or, which boils down to the same, in Fig. 3.3 we distinguish between two zigzag paths that are mirror images of each other with respect to the middle line of the figure (which joins the initial A to the final A). Or, in ex. 3.29, the r faces have a predetermined color different from that of the s faces, even if the numerical values of r and s happen to coincide.

3.31. This fact is accessible through all four approaches indicated in sect. 3.8 and also through ex. 3.23.

(1) The network of streets is symmetrical with respect to the vertical through the apex of the Pascal triangle.

(2) The same symmetry appears both in the recursion formula and in the boundary condition.

(3) Using the notation for the factorial

$$1 \cdot 2 \cdot 3 \ldots m = m!$$

we have

$$
\begin{aligned}
\binom{n}{r} &= \frac{n(n-1)\ldots(n-r+1)}{1 \cdot 2 \quad \ldots \quad r} \\
&= \frac{n(n-1)\ldots(n-r+1)(n-r)\ldots 2\cdot 1}{1 \cdot 2 \quad \ldots \quad r \quad (n-r)\ldots 2 \cdot 1} \\
&= \frac{n!}{r!(n-r)!} = \frac{n!}{(n-r)!r!} = \binom{n}{n-r}
\end{aligned}
$$

(4) Since $(a + b)^n$ remains unchanged when we interchange a and b, its expansion must show the same coefficient for $a^r b^{n-r}$ and $a^{n-r}b^r$.

(5) When, from a set of n individuals, we pick out a subset of r individuals, we leave another subset of $n - r$ individuals. Therefore, there are as many subsets of one kind as of the other kind.

3.32. $\quad \dbinom{n}{0} + \dbinom{n}{1} + \dbinom{n}{2} + \cdots + \dbinom{n}{n} = 2^n.$

Proof: put $a = b = 1$ in the expansion of $(a + b)^n$. *Another proof*: There are 2^n shortest zigzag paths from the apex of the Pascal triangle to its nth base; this is obvious since, in picking a southward path in Fig. 3.3, you have a choice between two alternatives in passing any street corner (any base). *Still another proof*: In a set of n individuals, there are 2^n subsets, including the empty set and the full set (which are accounted for by $\binom{n}{0}$ and $\binom{n}{n}$, respectively); this is obvious, since in picking a subset you may accept or refuse any one of the n individuals.

3.33.
$$\binom{n}{0} - \binom{n}{1} + \binom{n}{2} - \cdots + (-1)^n \binom{n}{n} = 0$$

for $n \geq 1$. Put $a = 1$ and $b = -1$ in the expansion of $(a + b)^n$.
Another proof: By boundary condition and recursion formula

$$\binom{n}{0} = \binom{n-1}{0}$$

$$-\binom{n}{1} = -\binom{n-1}{0} - \binom{n-1}{1}$$

$$\binom{n}{2} = \binom{n-1}{1} + \binom{n-1}{2}$$

$\cdots \cdots \cdots \cdots \cdots \cdots \cdots \cdots \cdots \cdots \cdots$

$$(-1)^{n-1}\binom{n}{n-1} = (-1)^{n-1}\binom{n-1}{n-2} + (-1)^{n-1}\binom{n-1}{n-1}$$

$$(-1)^n \binom{n}{n} = (-1)^n \binom{n-1}{n-1}$$

Add!
Still another proof: Each zigzag path attaining the $(n-1)$th base splits into two zigzag paths going to the nth base, of which one goes to a "positive" corner ($r = 0, 2, 4, \ldots$) and the other to a "negative" corner ($r = 1, 3, 5, \ldots$).

3.34. Analogously (fourth avenue)

$$1 + 5 + 15 + 35 = 56$$

generally (rth avenue)

$$\binom{r}{r} + \binom{r+1}{r} + \binom{r+2}{r} + \cdots + \binom{n}{r} = \binom{n+1}{r+1}$$

Proof by mathematical induction: The assertion is true for $n = r$: in fact,

$$\binom{r}{r} = \binom{r+1}{r+1}$$

by virtue of the boundary condition.

Assume now that the assertion holds for a certain value of n. Adding the same quantity to both sides of the assumed equation, we find

$$\binom{r}{r} + \binom{r+1}{r} + \cdots + \binom{n}{r} + \binom{n+1}{r} = \binom{n+1}{r+1} + \binom{n+1}{r} = \binom{n+2}{r+1}$$

by virtue of the recursion formula, and so the truth of the assertion follows for the next value, $n + 1$.

This proves the theorem for $n \geqq r$.

Another proof: In Fig. 3.7 (I), A is the apex and L a given point specified by $n + 1$ and $r + 1$; the total number of shortest zigzag paths from A to L is $\binom{n + 1}{r + 1}$. Each of these paths must use some street in going from the rth avenue to the $(r + 1)$th avenue; the number of paths using the successive streets is

$$\binom{r}{r}, \quad \binom{r + 1}{r}, \quad \binom{r + 2}{r}, \quad \ldots, \quad \binom{n}{r}$$

respectively, and so the sum of these numbers is the total number of the paths in question, $\binom{n + 1}{r + 1}$, as it has been asserted.

3.35. Adding the numbers first along the northwest boundary line (0th avenue), then along the first avenue, then along the second,... and finally along the fifth avenue in Fig. 3.5 we obtain

$$6, \quad 21, \quad 56, \quad 126, \quad 252, \quad 462$$

respectively, and the sum of these numbers is 923, which we seek in vain in the neighborhood of the fragment of the Pascal triangle exhibited in Fig. 3.5. Yet we find quite close the next number

$$924 = \binom{12}{6}$$

Observe now that we could have saved the trouble of performing our additions (also the last, the seventh addition) by using ex. 3.34 and a table of binomial coefficients, and you can easily prove, in ascending from our representative example, that generally

$$\sum_{l=0}^{m} \sum_{r=0}^{n} \binom{l + r}{r} = \binom{m + n + 2}{m + 1} - 1$$

3.36. On the left-hand side of the proposed equation, the first factors are taken from the fifth base and the second factors from the fourth base of the Pascal triangle; the right-hand side can be found in the ninth base. In the example $1 \cdot 1 + 5 \cdot 3 + 10 \cdot 3 + 10 \cdot 1 = 56$, the fifth, the third, and the eighth base are analogously involved. The more general situation considered in sect. 3.9 involves the nth, again the nth, and the $2n$th base. These examples suggest the general theorem:

$$\binom{m}{0}\binom{n}{r} + \binom{m}{1}\binom{n}{r - 1} + \binom{m}{2}\binom{n}{r - 2} + \cdots + \binom{m}{r}\binom{n}{0}$$

$$= \binom{m + n}{r}$$

We admit here, in fact, an extension of the meaning of our symbols; a formal statement follows in ex. 3.65 (III).

Both proofs found in sect. 3.9 can be extended to the present more general case. The geometric approach is suggested by the comparison of (II) and (III) in Fig. 3.7. The analytic approach consists in computing in two different ways the coefficient of x^r in the expansion of

$$(1 + x)^m(1 + x)^n = (1 + x)^{m+n}$$

3.37. On the left-hand side of the proposed equation, the first factors are taken from the first avenue and the second factors from the second avenue of the Pascal triangle; the right-hand side can be found on fourth avenue. In the example

$$1\cdot 10 + 3\cdot 6 + 6\cdot 3 + 10\cdot 1 = 56$$

the second, again the second, and the fifth avenue are analogously involved. We can interpret the general situation dealt with in ex. 3.34 and Fig. 3.7 (I) as involving the 0th, the rth and the $(r + 1)$th avenue in an analogous way. These examples suggest the general theorem:

$$\binom{r}{r}\binom{s + n}{s} + \binom{r + 1}{r}\binom{s + n - 1}{s}$$

$$+ \binom{r + 2}{r}\binom{s + n - 2}{s} + \cdots + \binom{r + n}{r}\binom{s}{s} = \binom{r + s + n + 1}{r + s + 1}$$

Geometric proof (more general than the geometric proof in ex. 3.34, analogous to that in sect. 3.9 and ex. 3.36): In Fig. 3.7(IV), the point L is specified by the numbers $r + 1 + s + n$ (total number of blocks) and $r + 1 + s$ (blocks to the right downward) and so the total number of shortest zigzag paths from the apex A to L is

$$\binom{r + s + n + 1}{r + s + 1}$$

Each of these paths must use some street in going from the rth avenue to the $(r + 1)$th avenue; according to the street used, we classify the paths on the left-hand side of the asserted formula, and count the paths of each class separately; on the right-hand side, we count all the paths in question together.

It would be desirable to parallel here also the other, analytic proof of sect. 3.9 and ex. 3.36 where the formula is derived from the consideration of the product of two series; yet this seems to be less immediate and there is a gap. It would be desirable too to find some (algebraic?) connection between the two similar formulas, obtained here and in the foregoing ex. 3.36; there is another gap.

3.38. The nth triangular number is

$$1 + 2 + 3 + \cdots + n = \frac{n(n + 1)}{2} = \binom{n + 1}{2}$$

The triangular numbers 1, 3, 6, 10,... form the second avenue of the Pascal triangle.

3.39. The nth pyramidal number is

$$\binom{2}{2} + \binom{3}{2} + \binom{4}{2} + \cdots + \binom{n+1}{2} = \binom{n+2}{3} = \frac{n(n+1)(n+2)}{6}$$

We have used ex. 3.34. The pyramidal numbers 1, 4, 10, 20,... form the third avenue of the Pascal triangle.

Remark. The expressions for the triangular and pyramidal numbers were known before the general explicit formula for the binomial coefficients (sect. 3.7) and may have led, by induction, to the discovery of the general formula.

3.40. $$1^2 + 2^2 + \cdots + n^2 = \frac{n(n+1)(2n+1)}{6}$$

3.41. Shortest zigzag paths joining the apex to the point characterized by the two numbers

$$n = n_1 + n_2 + \cdots + n_h$$

(total number of blocks) and

$$r = r_1 + r_2 + \cdots + r_h$$

(blocks from northwest to southeast) which, however, are subject to the *restriction* that they must pass through $h - 1$ given intermediate points, analogously characterized by the numbers

$$n_1 \qquad\qquad \text{and} \quad r_1$$
$$n_1 + n_2 \qquad\qquad \text{and} \quad r_1 + r_2$$
$$\cdots\cdots\cdots\cdots\cdots$$
$$n_1 + n_2 + \cdots + n_{h-1} \quad \text{and} \quad r_1 + r_2 + \cdots + r_{h-1}$$

3.42. (*a*) Fig. 3.10 represents *two* paths belonging to the set, but *not* to the subset (1). They have the same initial point A, the same final point C, and pass through the same intermediate point B which lies on the line of symmetry and cuts each path into two arcs, AB and BC. The arcs AB are symmetric to each other with respect to the line of symmetry and neither of them has an *interior* point in common with this line; the arcs BC coincide. Of these two paths, one belongs to the subset (2) and the other to the subset (3). Conversely, any path belonging to these subsets can be matched with another path in the manner presented by Fig. 3.10: look for the second common point B of the path with the line of symmetry (the apex A is the first such common point). Such matching establishes a one-to-one correspondence between the subsets (2) and (3).

(*b*) We could match the paths differently: whereas in Fig. 3.10 the two arcs AB are symmetrical to each other with respect to the straight line through the points A and B, in Fig. 3.11 they are symmetrical with respect to the midpoint of the segment AB.

(*c*) From (*a*) or (*b*) it follows that

$$\binom{n}{r} = N + 2\binom{n-1}{r}$$

Using first one then the other of the two following relations

$$\binom{n}{r} = \binom{n-1}{r-1} + \binom{n-1}{r}, \qquad \binom{n-1}{r} = \frac{n-r}{n}\binom{n}{r}$$

we obtain two different expressions

$$N = \binom{n-1}{r-1} - \binom{n-1}{r} = \frac{2r-n}{n}\binom{n}{r}$$

We have derived this in supposing that $2r > n$. Yet we can easily get rid of this restriction by using the symmetry of the Pascal triangle.

3.43. *With mathematical induction.* Verify the predicted result for $n = 1, 2, 3, (m = 0, 1)$ by inspecting the figure.

From $2m$ to $2m + 1$. By producing a path of length $2m$ which has no point in common with the line of symmetry except the apex, we obtain two paths of length $2m + 1$ of the same nature. Assuming the predicted result for $n = 2m$, we obtain so for $n = 2m + 1$

$$2\binom{2m}{m}$$

as the value of the required number.

From $2m + 1$ to $2m + 2$. By producing a path of length $2m + 1$ of the specified nature (see above) we obtain in most cases two paths of length $2m + 2$ of the same nature: the paths ending in the two points of the $(2m + 1)$th base nearest to the line of symmetry are exceptional. Visualizing this case, assuming the result for $n = 2m + 1$, and using the appropriate particular case of ex. 3.42, we obtain so, for $n = 2m + 2$, as the value of the required number

$$4\binom{2m}{m} - 2\frac{1}{2m+1}\binom{2m+1}{m+1}$$

which turns out, after suitable transformation, equal to

$$\binom{2m+2}{m+1}$$

Without mathematical induction. Use the first expression obtained for N in the solution of ex. 3.42 under (c), and extend the following sum to values of r restricted by $n/2 < r \le n$:

$$2\sum\left[\binom{n-1}{r-1} - \binom{n-1}{r}\right]$$

is the value of the required number and yields the predicted result, if we carefully distinguish between the cases $n = 2m$ and $n = 2m + 1$.

3.44.   0,   1,   6,   21,   50,   90,   126,   141,   126...
              0,   1,   7,   28,   77,   161,   266,   357,   393,   357...

1, 393, and 1 are not, the other numbers of the seventh base are, divisible by 7.

3.45. Analogous to ex. 3.1.

3.46. Analogous to ex. 3.31.

3.47. Analogous to ex. 3.32.

3.48. Analogous to ex. 3.33.

3.49. Analogous to sect. 3.9; a wider generalization is analogous to ex. 3.36.

3.50. The lines sloping from northeast to southwest

$$
\begin{array}{ccccc}
1, & 1, & 1, & 1, & 1, & \ldots \\
1, & 2, & 3, & 4, & 5, & \ldots \\
1, & 3, & 6, & 10, & 15, & \ldots
\end{array}
$$

are also "avenues" in the Pascal triangle.

3.51. The symmetry, visible in the first lines, persists; it is enough to write out two bases (the seventh and the eighth) to the middle

$$
\frac{1}{8} \qquad \frac{1}{56} \qquad \frac{1}{168} \qquad \frac{1}{280}
$$

$$
\frac{1}{9} \qquad \frac{1}{72} \qquad \frac{1}{252} \qquad \frac{1}{504} \qquad \frac{1}{630}
$$

3.52. In a given "base" of the harmonic triangle, the denominators are proportional to the binomial coefficients, and the factor of proportionality is visible from the extreme terms. More explicitly, we find in corresponding location in the two triangles the numbers

$$
\binom{n}{r} \qquad \frac{1}{(n+1)\binom{n}{r}}
$$

Pascal   Leibnitz

Proof. For $r = 0$, the boundary condition of the harmonic triangle is verified. To verify its recursion formula, use first the recursion formula, then the explicit form, of the binomial coefficients:

$$
\frac{1}{(n+1)\binom{n}{r-1}} + \frac{1}{(n+1)\binom{n}{r}} = \frac{\binom{n+1}{r}}{(n+1)\binom{n}{r-1}\binom{n}{r}}
$$

$$
= \frac{1}{n+1} \cdot \frac{(n+1)!}{r!(n+1-r)!} \cdot \frac{(r-1)!(n-r+1)!}{n!} \cdot \frac{r!(n-r)!}{n!}
$$

$$
= \frac{(r-1)!(n-r)!}{n!} = \frac{1}{n\binom{n-1}{r-1}}
$$

3.53. On the left-hand side, there is the initial term of an avenue in Fig. 3.13, and on the right-hand side the sum of all terms in the next avenue. For a proof, see the solution of ex. 3.54.

3.54. Use the recursion formula of the Leibnitz triangle:

$$\frac{1}{6} - \frac{1}{12} = \frac{1}{12}$$

$$\frac{1}{12} - \frac{1}{20} = \frac{1}{30}$$

$$\frac{1}{20} - \frac{1}{30} = \frac{1}{60}$$

$$\frac{1}{30} - \frac{1}{42} = \frac{1}{105}$$

. . . . . .

Add ! (A "faraway" term of the second avenue is "negligible.") From this representative particular case, we easily pass to the general proposition: In the Leibnitz triangle, the sum of all the (infinitely many) terms of the avenue beginning with, and to the southwest from, a certain initial term, is the northwest neighbor of the initial term. By changing

"Leibnitz"   "infinitely many"   "southwest"   "northwest"

into

"Pascal"   "finite number of"   "northeast"   "southeast"

we pass from the present result to that of ex. 3.34, in which we may see a further manifestation of that "analogy by contrast" observed in ex. 3.51.

3.55. In view of the explicit formula for the general term of the Harmonic Triangle (ex. 3.52) the displayed $(r - 1)$th line differs only in a factor from the corresponding line of ex. 3.53, for $r = 2, 3, \ldots$, and its sum is

$$\frac{1}{(r - 1)!(r - 1)}$$

3.57. The product is $= 1$. The reader acquainted with the theory of infinite series understands the equation

$$1 + x + x^2 + \cdots + x^n + \cdots = (1 - x)^{-1}$$

in a less formal meaning, knows the condition under which it possesses that meaning, and knows also a satisfactory derivation.

3.58. $a_0 + a_1 + a_2 + \cdots + a_n$; ex. 3.57 is a particular case.

3.59. Each series corresponds to an avenue of the Pascal triangle. For the initial series see ex. 3.57. By repeated application of ex. 3.58 and ex. 3.34 we find that

$$1 + 2x + 3x^2 + 4x^3 + \cdots = (1 + x + x^2 + \cdots)^2$$
$$1 + 3x + 6x^2 + 10x^3 + \cdots = (1 + x + x^2 + \cdots)^3$$

and, generally,

$$\binom{r}{r} + \binom{r + 1}{r}x + \binom{r + 2}{r}x^2 + \cdots + \binom{r + n}{r}x^n + \cdots$$

$$= (1 + x + x^2 + \cdots)^{r+1} = (1 - x)^{-r-1}$$

For a formal proof, use mathematical induction.

3.60. Compute in two different ways the coefficient of x^n in the product

$$(1 - x)^{-r-1}(1 - x)^{-s-1}$$

This is strictly analogous to the analytic approach in the solution of ex. 3.36, which goes back to sect. 3.9(3).

3.61. 1, 0, 0, 0, respectively, which can be regarded as a confirmation of the conjecture N.

3.62. $\dfrac{2}{3}$, $-\dfrac{1}{9}$, $\dfrac{4}{81}$, $-\dfrac{7}{243}$ respectively, computed by two essentially different

procedures, which can be regarded as another confirmation of the conjecture N.

3.63. $(1 + x)^{1/3}(1 + x)^{2/3}$

$$= \left(1 + \frac{x}{3} - \frac{x^2}{9} + \frac{5x^3}{81} - \frac{10x^4}{243} + \cdots\right)\left(1 + \frac{2x}{3} - \frac{x^2}{9} + \frac{4x^3}{81} - \frac{7x^4}{243} + \cdots\right)$$

$$= 1 + x + 0x^2 + 0x^3 + 0x^4 + \cdots$$

which yields a further confirmation for the conjecture N.

3.64.

$$1 + \frac{-1}{1} x + \frac{(-1)(-2)}{1 \cdot 2} x^2 + \frac{(-1)(-2)(-3)}{1 \cdot 2 \cdot 3} x^3 + \cdots$$

$$= 1 - x + x^2 - x^3 + \cdots$$

$$= [1 - (-x)]^{-1} = (1 + x)^{-1}$$

by virtue of ex. 3.57, which confirms conjecture N from a quite different side. Can the other series of ex. 3.59 also be derived from conjecture N?

3.65. $\dbinom{r - 1 - x}{r} = \dfrac{r - 1 - x}{1} \cdot \dfrac{r - 2 - x}{2} \cdots \dfrac{-x}{r}$

$$= (-1)^r \frac{x}{1} \cdots \frac{x - r + 2}{r - 1} \cdot \frac{x - r + 1}{r}$$

3.66. According to conjecture N, the coefficient of x^n in the expansion of

$(1 + x)^{-r-1}$ is $\dbinom{-r - 1}{n} = (-1)^n\dbinom{n + r}{n} = (-1)^n\dbinom{r + n}{r}$; we have first

used ex. 3.65 (II), and then we have supposed that r *is a non-negative integer* and used ex. 3.31. Replacing x by $-x$, and so x^n by $(-1)^n x^n$, we obtain the general result of ex. 3.59, which proves the conjecture N in an extensive particular case: for negative integral values of a.

3.67. From

$$\left[\binom{a}{0} + \binom{a}{1}x + \cdots + \binom{a}{r}x^r + \cdots\right]$$

$$\times \left[\binom{b}{0} + \cdots + \binom{b}{r - 1}x^{r-1} + \binom{b}{r}x^r + \cdots\right]$$

$$= \binom{a + b}{0} + \binom{a + b}{1}x + \cdots + \binom{a + b}{r}x^r + \cdots$$

we infer (ex. 3.56) that

$$(*) \quad \binom{a}{0}\binom{b}{r} + \binom{a}{1}\binom{b}{r-1} + \cdots + \binom{a}{r}\binom{b}{0} = \binom{a+b}{r}$$

If we put $a = m$ and $b = n$, this goes over into the result of ex. 3.36, but the range is different: m and n are restricted to be non-negative integers, a and b are unrestricted, arbitrary numbers.

3.68. Relation $(*)$, derived from the conjecture N, is not proved: it is just a conjecture.

The particular case of $(*)$ in which a and b are positive integers has been proved in ex. 3.36. In fact, in view of the solution of ex. 3.66, the particular case of $(*)$ in which a and b are negative integers is equivalent to the result of ex. 3.57, and so it is also proved. (Observe that $(*)$ provides so the desired connection between ex. 3.36 and ex. 3.37; see the remark at the end of the solution of ex. 3.37.)

Could we use ex. 3.36, which is an extensive particular case of the desired $(*)$, as a stepping stone to prove the full statement $(*)$? (Yes, we can, if we know the relevant algebraic facts: a polynomial in two variables x and y must vanish identically if it vanishes for all positive integral values of x and y.)

Set

$$\binom{a}{0} + \binom{a}{1}x + \binom{a}{2}x^2 + \cdots + \binom{a}{n}x^n + \cdots = f_a(x)$$

The relation $(*)$ is essentially equivalent to the relation

$$f_a(x)f_b(x) = f_{a+b}(x)$$

Now, take $(*)$ for granted; there follows

$$f_a(x)f_a(x)f_a(x) = f_{2a}(x)f_a(x) = f_{3a}(x)$$

and, generally,

$$f_a(x)^n = f_{na}(x)$$

for any positive integer n. Let m be a (positive or negative) integer; since we have verified already conjecture N for positive and negative integral values of a (see ex. 3.1 and ex. 3.66, respectively) we infer that

$$[f_{m/n}(x)]^n = f_m(x) = (1 + x)^m$$

$$f_{m/n}(x) = (1 + x)^{m/n}$$

and so we have *derived from* $(*)$ *the conjecture N for all rational values of the exponent a.*

(In fact, the last step is rather risky: in extracting the nth root, we failed to indicate which one of its possible values is meant, and thus we left a gap which we can hardly fill if we remain on the purely formal standpoint of ex. 3.56. Still, we have discovered essential materials for the construction of a full proof. One century and a half after Newton's letter, in 1826, there appeared a memoir of the great Norwegian mathematician Niels Henrik Abel in which he discussed the convergence and the value of the binomial series, also for complex values of x and a, and greatly advanced the general theory of infinite series; see his *Œuvres complètes*, 1881, vol. 1, p. 219–250.)

3.69. We find 1, 2, 6, 20 on the line of symmetry of the Pascal triangle. Explanation: the coefficient of x^n is

$$(-4)^n\binom{-\frac12}{n} = 4^n\frac{1\cdot3\cdot5\ldots(2n-1)}{2\cdot4\cdot6\ldots\ 2n}$$
$$= \frac{1\cdot3\cdot5\ldots(2n-1)\cdot2\cdot4\cdot6\ldots2n}{n!n!}$$
$$= \binom{2n}{n}$$

3.70.

$$a_0u_0 = b_0$$
$$a_0^2u_1 = a_0b_1 - a_1b_0$$
$$a_0^3u_2 = a_0^2b_2 - a_0a_1b_1 + (a_1^2 - a_0a_2)b_0$$
$$a_0^4u_3 = a_0^3b_3 - a_0^2a_1b_2 + (a_0a_1^2 - a_0^2a_2)b_1 - (a_1^3 - 2a_0a_1a_2 + a_0^2a_3)b_0$$

3.71. The cases $n = 0, 1, 2, 3$ treated in ex. 3.70 suggest that $a_0^{n+1}u_n$ is a polynomial in the a's and b's all terms of which have

(1) the same degree n in the a's
(2) the same degree 1 in the b's
(3) the same weight n in the a's and b's jointly.

Reasons:

(1) If a_n is replaced by a_nc (for $n = 0, 1, 2,\ldots$, with arbitrary c) u_n must be replaced by u_nc^{-1}.
(2) If b_n is replaced by b_nc, u_n must be replaced by u_nc.
(3) If a_n and b_n are replaced by a_nc^n and b_nc^n respectively (as a result of substituting cx for x) also u_n must be replaced by u_nc^n.

3.72. $u_n = b_n - b_{n-1}$: this value must result if, having expressed u_n in terms of a's and b's, we set $a_0 = a_1 = a_2 = a_3 = \cdots = 1$. This is a valuable check; carry it through for $n = 0, 1, 2, 3$ (ex. 3.70).

3.73. $u_n = b_0 + b_1 + b_2 + \cdots + b_n$ (see ex. 3.58): this value must result if, having expressed u_n in terms of a's and b's, we set $a_0 = 1$, $a_1 = -1$, $a_2 = a_3 = \cdots = 0$. This is a valuable check; carry it through for $n = 0, 1, 2, 3$ (ex. 3.70.)

3.74.

$$1 - \frac{x}{6} + \frac{x^2}{40} - \frac{x^3}{336} + \cdots + \frac{(-1)^nx^n}{(2\cdot4\cdot6\ldots2n)(2n+1)} + \cdots$$

Cf. MPR, vol. 1, p. 84, ex. 2.

3.75.

$$a_1u_1 = 1$$
$$-a_1^3u_2 = a_2$$
$$a_1^5u_3 = 2a_2^2 - a_1a_3$$
$$-a_1^7u_4 = 5a_2^3 - 5a_1a_2a_3 + a_1^2a_4$$
$$a_1^9u_5 = 14a_2^4 - 21a_1a_2^2a_3 + 3a_1^2a_3^2 + 6a_1^2a_2a_4 - a_1^3a_5$$

3.76. The cases treated in ex. 3.75 suggest that $a_1{}^{2n-1}u_n$ is a polynomial in the a's each term of which is

(1) of degree $n - 1$ and
(2) of weight $2n - 2$.

Reasons:

(1) If a_n is replaced by $a_n c$ (as a result of substituting $c^{-1}x$ for x) u_n must be replaced by $u_n c^{-n}$.

(2) If a_n is replaced by $a_n c^n$ (as a result of substituting cy for y) u_n must be replaced by $u_n c^{-1}$.

3.77. $x = \dfrac{y}{1 - y}$, $y = \dfrac{x}{1 + x}$ and so

$$y = x - x^2 + x^3 - x^4 + \cdots$$

Hence, if we set $a_n = 1$, in ex. 3.75, we must get $u_n = (-1)^{n-1}$.   This is a valuable check; carry it through for $n = 1, 2, 3, 4, 5$.

3.78. $1 - 4x = (1 + y)^{-2}$ or

$$y = -1 + (1 - 4x)^{-\frac{1}{2}} = 2x + 6x^2 + \cdots + \binom{2n}{n}x^n + \cdots$$

see ex. 3.69.

3.79. $y = -1 + (1 + 4ax)^{\frac{1}{2}}(2a)^{-1}$
$\qquad = x - ax^2 + 2a^2x^3 - 5a^3x^4 + 14a^4x^5 - \cdots$

The coefficient of x^n is

$$\frac{(4a)^n}{2a}\binom{\frac{1}{2}}{n} = \frac{(-1)^{n-1}a^{n-1}}{n}\binom{2n-2}{n-1}$$

(computation similar to ex. 3.69) and u_n of ex. 3.75 must reduce to this value if $a_1 = 1$, $a_2 = a$, $a_3 = a_4 = \cdots = 0$.   Cf. MPR, vol. 1, p. 102, ex. 7, 8, 9.

3.80.

$$y = x - \frac{x^2}{2} + \frac{x^3}{3} - \frac{x^4}{4} + \cdots + \frac{(-1)^{n-1}x^n}{n} + \cdots$$

3.81. $u_0 = u_1 = u_2 = 1$, $u_3 = \dfrac{4}{3}$, $u_4 = \dfrac{7}{6}$.

3.82. Mathematical induction: The assertion is true for $n = 3$.   Assume that $n > 3$ and that the assertion has been proved for the coefficients preceding u_n so that

$$u_{n-1} > 1, \quad u_{n-2} > 1, \ldots, u_3 > 1$$

We know that $u_0 = u_1 = u_2 = 1$ and, therefore,

$$nu_n = u_0 u_{n-1} + u_1 u_{n-2} + \cdots + u_{n-1}u_0 > n$$

3.83. Set

$$y = u_0 + u_1 x + u_2 x^2 + \cdots + u_n x^n + \cdots$$

$$\frac{d^2y}{dx^2} = \qquad\qquad 2 \cdot 1 u_2 + \cdots + n(n-1)u_n x^{n-2} + \cdots$$

From the differential equation

$$n(n - 1)u_n = -u_{n-2}$$

From the initial condition

$$u_0 = 1, \qquad u_1 = 0$$

Finally, for $m = 1, 2, 3, \ldots,$

$$u_{2m} = \frac{(-1)^m}{2m!}, \qquad u_{2m-1} = 0$$

$$y = 1 - \frac{x^2}{2!} + \frac{x^4}{4!} - \frac{x^6}{6!} + \cdots$$

3.84. $B_n = B_{n-5} + A_n$
$C_n = C_{n-10} + B_n$
$D_n = D_{n-25} + C_n$
$E_n = E_{n-50} + D_n$

The last equation yields, for $n = 100,$

$$E_{100} = E_{50} + D_{100}$$

and the foregoing equation yields for $n = 20,$

$$D_{20} = C_{20}$$

since $D_{-5} = 0$: any of the introduced quantities with a negative subscript must be properly regarded as equal to 0.   These examples should illustrate the main feature of the system of equations obtained: we can compute any unknown (such as E_{100}) if a certain unknown of the same kind with a lower subscript (as E_{50}) and another unknown of a lower kind (denoted by the foregoing letter of the alphabet, as D_{100}) have been computed previously.   (There are cases in which only one previously computed unknown is needed, see D_{20}.   In other cases we need some of the "boundary values" which we have known before setting up our equations: I mean $B_0, C_0, D_0, E_0,$ and A_n for $n = 0, 1, 2, 3, \ldots$)   In short, we compute the unknowns by going back to lower subscripts or to former letters of the alphabet and, eventually, to the boundary values. (Do not let the difference of notation hide the analogy between this computation and the determination of the binomial coefficients by recursion formula and boundary condition; see sect. 3.6(2).)

The reader should set up a convenient system of computation which he can check against the following values:

$$B_{10} = 3, \qquad C_{25} = 12, \qquad D_{50} = 49, \qquad E_{100} = 292$$

(For more details, and a concrete interpretation of the problem, see HSI, p. 238, ex. 20, and *American Mathematical Monthly*, **63**, 1956, pp. 689–697.)

3.85. Mathematical induction: assuming the proposed form for $y^{(n)}$, we find by differentiating once more

$$y^{(n+1)} = (-1)^{n+1}(n + 1)!x^{-n-2} \log x$$
$$+ (-1)^n x^{-n-2}[n! + (n + 1)c_n]$$

which is of the desired form provided that

$$c_{n+1} = n! + (n+1)c_n$$

Transforming this into

$$\frac{c_{n+1}}{(n+1)!} = \frac{c_n}{n!} + \frac{1}{n+1}$$

and using $c_1 = 1$, we find

$$c_n = n!\left(1 + \frac{1}{2} + \frac{1}{3} + \frac{1}{4} + \cdots + \frac{1}{n}\right)$$

3.86. To find the sum of a geometric series is a closely related problem: both the result and the usual method of derivation are used in the following.

Call S the proposed sum. Then

$$(1 - x)S = 1 + x + x^2 + \cdots + x^{n-1} - nx^n$$
$$= \frac{1 - x^n}{1 - x} - nx^n$$

Hence the required short expression:

$$S = \frac{1 - (n+1)x^n + nx^{n+1}}{(1 - x)^2}$$

3.87. Use method, notation, and result of ex. 3.86: call T the proposed sum and consider

$$(1 - x)T = 1 + 3x + 5x^2 + 7x^3 + \cdots + (2n - 1)x^{n-1} - n^2x^n$$
$$= 2S - (1 + x + x^2 + \cdots + x^{n-1}) - n^2x^n$$

and so by straightforward algebra

$$T = \frac{1 + x - (n+1)^2x^n + (2n^2 + 2n - 1)x^{n+1} - n^2x^{n+2}}{(1 - x)^3}$$

3.88. In following ex. 3.86 and 3.87, we could find an expression for

$$1^k + 2^kx + 3^kx^2 + \cdots + n^kx^{n-1}$$

by recursion, by reducing the case k to the cases $k - 1, k - 2, \ldots, 2, 1, 0$.

3.89. Mathematical induction. The assertion is obviously true for $n = 1$, and

$$\frac{a_n(n + \alpha) - a_1(1 + \beta)}{\alpha - \beta} + a_{n+1} = \frac{a_{n+1}(n + 1 + \beta) - a_1(1 + \beta)}{\alpha - \beta} + a_{n+1}$$

3.90. Apply ex. 3.89 with

$$a_1 = \frac{p}{q}, \qquad \alpha = p, \qquad \beta = q - 1$$

The proposed sum turns out to equal

$$\frac{p}{p - q + 1}\left(\frac{p+1}{q}\frac{p+2}{q+1}\cdots\frac{p+n}{q+n-1} - 1\right)$$

3.91. (1) $8, 4\sqrt{2}, 4\sqrt{3}, 6$.

(2) $C_n = 2n \tan (\pi/n)$, $I_n = 2n \sin (\pi/n)$.

Hence straightforward verification by familiar trigonometric identities.

3.92. For more examples on mathematical induction see p.75 footnote 5. Problems connected with those in Parts II and III can be found in books on Calculus of Probability or Combinatorial Analysis. Problems connected with those in Part IV, or with ex. 3.53 to 3.55, can be found in books on Infinite Series or Complex Variables. Problems closely related to ex. 3.81, 3.82 and 3.83 fill an extensive chapter of the Theory of Differential Equations.

There is an inexhaustible source of themes for further examples. Here is just one instance: *polynomial coefficients* (cf. ex. 3.28, 3.29, 3.30). The coefficients of the expansion of

$$(a + b + c)^n$$

for $n = 0, 1, 2, 3, \ldots$ can be associated with the lattice points in an octant of space analogously to the coefficients of

$$(a + b)^n$$

which are, in the Pascal triangle, associated with the lattice points in one quarter of a plane. What is, in the space arrangement, analogous to the boundary condition, to the recursion formula, to the avenues, streets, and bases of the Pascal triangle, to ex. 3.31–3.39? What is the connection with ex. 3.44–3.50? We have not yet mentioned the number-theoretic properties of binomial, or polynomial, coefficients. And so on.

No solution: **3.56.**

Chapter 4

4.1. Let A denote the vertex of the pyramid opposite its base (the "apex"). Dissect the base of the pyramid into n triangles of area

$$B_1, B_2, \ldots, B_n$$

respectively. Each of these triangles forms the base of a tetrahedron of which A is the opposite vertex and h the height; the pyramid is dissected (by planes passing through A) into these n tetrahedra of volume

$$V_1, V_2, \ldots, V_n$$

respectively. Obviously,

$$B_1 + B_2 + \cdots + B_n = B$$
$$V_1 + V_2 + \cdots + V_n = V$$

Supposing that the desired expression for the volume has been proved for the particular case of the tetrahedra, we have

$$V_1 = \frac{B_1 h}{3}, \quad V_2 = \frac{B_2 h}{3}, \quad \ldots, \quad V_n = \frac{B_n h}{3}$$

Addition (superposition!) of these special relations yields the general relation

$$V = \frac{Bh}{3}$$

4.2. A polynomial of degree k is of the form

$$f(x) = a_0 x^k + a_1 x^{k-1} + \cdots + a_k$$

where $a_0 \neq 0$. Substitute successively $1, 2, 3, \ldots, n$ for x and add: you obtain, in using the notation of ex. 3.11,

$$f(1) + f(2) + \cdots + f(n) = a_0 S_k(n) + a_1 S_{k-1}(n) + \cdots + a_k S_0(n)$$

The right-hand side is, by the result of ex. 3.3, a polynomial of degree $k + 1$ in n.

4.3. We can write the result of ex. 3.34 in the form

$$\binom{0}{r} + \binom{1}{r} + \binom{2}{r} + \cdots + \binom{n}{r} = \binom{n+1}{r+1}$$

see ex. 3.65(III). Assuming ex. 4.4, we write the polynomial considered in the form

$$f(x) = b_0 \binom{x}{k} + b_1 \binom{x}{k-1} + \cdots + b_k \binom{x}{0}$$

where (see the solution of ex. 4.4) $b_0 = k! a_0 \neq 0$. Substitute successively $0, 1, 2, 3, \ldots, n$ for x and add: you obtain, by the result of ex. 3.34 restated above,

$$f(0) + f(1) + f(2) + \cdots + f(n) = b_0 \binom{n+1}{k+1} + b_1 \binom{n+1}{k} + \cdots + b_k \binom{n+1}{1}$$

The right-hand side is a polynomial of degree $k + 1$ in n.

4.4. Comparing the coefficient of x^k (the highest power of x present) on both sides of the proposed identity, we find

$$a_0 = b_0/k!$$

Hence, it follows from the proposed identity

$$a_0 x^k + a_1 x^{k-1} + \cdots + a_k - k! a_0 \binom{x}{k} = b_1 \binom{x}{k-1} + \cdots + b_k \binom{x}{0}$$

Comparing the coefficient of x^{k-1} on both sides, we express b_1 in terms of a_0 and a_1, and in continuing in this fashion we determine $b_0, b_1, b_2, \ldots, b_k$ successively, by recursion.

4.5. We have to determine four numbers b_0, b_1, b_2 and b_3 so that

$$x^3 = b_0 \binom{x}{3} + b_1 \binom{x}{2} + b_2 \binom{x}{1} + b_3 \binom{x}{0}$$

holds identically in x. This means

$$x^3 = \frac{b_0}{6} (x^3 - 3x^2 + 2x) + \frac{b_1}{2} (x^2 - x) + b_2 x + b_3$$

Comparison of the coefficients of x^3, x^2, x^1, and x^0 yields the equations

$$1 = b_0/6$$
$$0 = -b_0/2 + b_1/2$$
$$0 = b_0/3 - b_1/2 + b_2$$
$$0 = b_3$$

respectively, from which we derive

$$b_0 = 6, \qquad b_1 = 6, \qquad b_2 = 1, \qquad b_3 = 0$$

Hence, by the procedure of ex. 4.3 ($k = 3$)

$$1^3 + 2^3 + \cdots + n^3 = 6\binom{n+1}{4} + 6\binom{n+1}{3} + \binom{n+1}{2}$$

$$= \frac{(n+1)^2 n^2}{4}$$

by straightforward algebra.

4.6. As it has been proved in ex. 4.3 there are five constants c_0, c_1, c_2, c_3, and c_4 such that

$$1^3 + 2^3 + 3^3 + \cdots + n^3 = c_0 n^4 + c_1 n^3 + c_2 n^2 + c_3 n + c_4$$

for all positive integral values of n. In setting successively $n = 1, 2, 3, 4,$ and 5, we obtain a system of five equations for the five unknowns c_0, c_1, c_2, c_3, and c_4. By solving these equations we obtain

$$c_0 = 1/4, \qquad c_1 = 1/2, \qquad c_2 = 1/4, \qquad c_3 = 0, \qquad c_4 = 0$$

that is, we obtain the same result as in ex. 4.5, but with more trouble.

4.7. Ex. 4.3 yields a new proof for the result of ex. 3.3, except one point: the coefficient of n^{k+1} in the expression of $S_k(n)$ remains undetermined by the procedure of ex. 4.3. (A little additional remark, however, will yield also this coefficient.)

4.8. Yes: a straight line is represented by an equation of the form

$$y = ax + b$$

the right-hand side of which is a polynomial of degree ≤ 1.

4.9. The straight line coinciding with the x axis appears intuitively as the simplest interpolating curve; it corresponds to the identically vanishing polynomial. Any *other* interpolating polynomial is necessarily of higher degree, namely of degree n at least, since it has n different zeros $x_1, x_2, \ldots, x_n$.

4.10. Lagrange's interpolating polynomial, given by the final formula of sect. 4.3, is of degree $\leq n - 1$; it is the only interpolating polynomial of such a low degree, I say. In fact, if two polynomials, both of degree $\leq n - 1$, take the same values at the n given abscissas, their difference has n different zeros, that is more zeros than the degree would permit, unless this difference vanishes identically. Lagrange's interpolating polynomial, being the only one of degree $\leq n - 1$, *is* of the lowest possible degree.

4.12. (*a*) Obvious, in view of the rule

$$(c_1 y_1 + c_2 y_2)' = c_1 y_1' + c_2 y_2'$$

for constant c_1 and c_2.

(*b*) $y = e^{rz}$ is a solution of the differential equation iff r is a root of the algebraic equation in r

$$r^n + a_1 r^{n-1} + a_2 r^{n-2} + \cdots + a_n = 0$$

(*c*) If the equation in r under (*b*) has d *different* roots $r_1, r_2, \ldots, r_d$ and $c_1, c_2, \ldots, c_d$ are arbitrary constants

$$y = c_1 e^{r_1 z} + c_2 e^{r_2 z} + \cdots + c_d e^{r_d z}$$

is a solution of the differential equation, and its most general solution (as it can be shown) when $d = n$.

4.13. The equation in r is

$$r^2 + 1 = 0$$

and so the general solution of the differential equation is

$$y = c_1 e^{iz} + c_2 e^{-iz}$$

The initial conditions yield the equations

$$c_1 + c_2 = 1, \qquad ic_1 - ic_2 = 0$$

which determine c_1 and c_2 and so the desired particular solution

$$y = (e^{iz} + e^{-iz})/2$$

Observe that also $y = \cos x$ satisfies both the differential equation and the boundary conditions. See also ex. 3.83.

4.14. (*a*) Obvious.

(*b*) $y_k = r^k$ is a solution of the difference equation iff r is a root of the algebraic equation given in ex. 4.12(*b*).

(*c*) If the equation of ex. 4.12(*b*) has d *different* roots $r_1, r_2, \ldots, r_d$ and $c_1, c_2, \ldots, c_d$ are arbitrary constants

$$y_k = c_1 r_1^k + c_2 r_2^k + \cdots + c_d r_d^k$$

is a solution of the difference equation and its most general solution (as it can be shown) when $d = n$.

4.15. The equation in r is

$$r^2 - r - 1 = 0$$

and so the general solution of the difference equation is

$$y_k = c_1 \left(\frac{1 + \sqrt{5}}{2} \right)^k + c_2 \left(\frac{1 - \sqrt{5}}{2} \right)^k$$

This yields for $k = 0$ and $k = 1$ (initial conditions) the equations

$$c_1 + c_2 = 0, \qquad c_1 \frac{1 + \sqrt{5}}{2} + c_2 \frac{1 - \sqrt{5}}{2} = 1$$

which determine c_1 and c_2 and so the desired expression of the Fibonacci numbers

$$y_k = \frac{1}{\sqrt{5}} \left[\left(\frac{1 + \sqrt{5}}{2} \right)^k - \left(\frac{1 - \sqrt{5}}{2} \right)^k \right]$$

4.16. If the actual motion is obtainable by *superposition* of the three virtual motions, the coordinates of the moving point at the time t are

$$x = x_1 + x_2 + x_3 = tv \cos \alpha,$$
$$y = y_1 + y_2 + y_3 = tv \sin \alpha - \tfrac{1}{2}gt^2$$

Elimination of t yields the trajectory of the projectile

$$y = x \tan \alpha - \frac{gx^2}{2v^2 \cos^2 \alpha}$$

which is a *parabola*.

4.18. There are two unknowns: the base and the height of the tetrahedron. See Fig. 4.5a.

4.19. Let

   V stand for the volume of the tetrahedron,
   B for its base,
   H for its height,
   h for that height of its base that is perpendicular to the edge of given
      length a.

Then

$$V = \frac{BH}{3}, \qquad B = \frac{ah}{2}$$

and, therefore,

$$V = \frac{ahH}{6}$$

Yet neither h nor H is given.

4.20. The orthogonal projection of our tetrahedron (of ex. 4.17) on a plane, perpendicular to the line of length b and passing through one of its endpoints, is a square. The diagonals of this square are of length a, its area is $a^2/2$, and the square itself is, see Fig. 4.5b, the base of a right prism with height b. This prism, see Fig. 4.5b, is split into five (nonoverlapping) tetrahedra; our tetrahedron of ex. 4.17 is one of them (we call its volume V); the other four are congruent, the base of each is an isosceles right triangle with area $a^2/4$, and the height of each is b. Hence

$$a^2 b/2 = V + 4a^2 b/12$$
$$V = a^2 b/6$$

4.21. The plane that passes through an edge of length a and the midpoint of the opposite edge is a plane of symmetry for our tetrahedron and divides it into two congruent tetrahedra (see Fig. 4.5c) the common base of which (an isosceles triangle) has obviously the area $ab/2$ and the height of which is $a/2$. Hence the desired volume

$$V = \frac{2}{3} \cdot \frac{a}{2} \cdot \frac{ab}{2} = \frac{a^2 b}{6}$$

(There are two such planes of symmetry which jointly divide our tetrahedron into four congruent tetrahedra; this yields another, but little different, access to the solution.)

4.22. We can regard our tetrahedron as an extreme (degenerate, limiting) prismoid, with height b and each base shrunk into a line-segment of length a; the midsection is a square with side $a/2$; see Fig. 4.5d. Hence

$$h = b, \qquad L = 0, \qquad M = a^2/4, \qquad N = 0$$

and the prismoidal formula yields $V = a^2 b/6$.

4.23. If the expression found for V in ex. 4.19 agrees with the result derived in three different ways in ex. 4.20, 4.21, and 4.22, we should have

$$Hh = ab$$

Yet we can show this relation independently, by computing, in two different ways, the area of the isosceles triangle, in which the tetrahedron is intersected by a plane of symmetry (ex. 4.21, Fig. 4.5e). And so we brought to a successful end a fourth (somewhat tortuous) derivation started in ex. 4.18 and continued through ex. 4.19.

4.24. The route from ex. 4.18 through ex. 4.19 to ex. 4.23 is too long and tortuous. The solution in ex. 4.21 appears the most elegant: it exploits fully the symmetry of the figure—but just for this reason it may be less applicable to nonsymmetric cases. Thus, *prima facie* evidence favors ex. 4.20. Do you see another indication in favor of ex. 4.20?

4.25. $L = M = N$, and so $V = Lh$.

4.26. $N = 0$, $M = L/4$, and so $V = Lh/3$.

4.27. Let L_i, M_i, N_i, and V_i denote the quantities that are so related to P_i as L, M, N, and V are to P respectively, for $i = 1, 2, \ldots, n$. All prismoids have the same height h. Obviously

$$L_1 + L_2 + \cdots + L_n = L$$
$$M_1 + M_2 + \cdots + M_n = M$$
$$N_1 + N_2 + \cdots + N_n = N$$
$$V_1 + V_2 + \cdots + V_n = V$$

By combining these equations we obtain

$$\sum_{i=1}^{n} \left(\frac{L_i + 4M_i + N_i}{6} h - V_i \right) = \frac{L + 4M + N}{6} h - V$$

We regard the right-hand side as *one* term; the left-hand side is a sum of n analogous terms. If n terms out of these $n + 1$ terms linked by our equation vanish, the remaining one term must vanish too.

4.28. The orthogonal projection of our tetrahedron onto the plane we have passed through *l* is a quadrilateral (a square in the particular case of ex. 4.20, Fig. 4.5*b*, but irregular in general).   One diagonal of this quadrilateral is the edge *l*, the other diagonal is parallel and equal to *n*.   This quadrilateral is the base of a prism with height *h*; the prism is split into five tetrahedra; one of them is our tetrahedron, the other four are pyramids in the situation described by ex. 4.26, and so the prismoidal formula is valid for them.   This formula is also valid for the prism, by ex. 4.25, and so also for our tetrahedron, by ex. 4.27.

4.29. Fig. S4.29 shows a prismoid; *B, C, . . . , K* are the vertices of the lower base (in the plane of the paper), and *B′, C′, . . . , K′* are the vertices of the upper base.

(1) Consider the pyramid of which the base is the upper base of the prismoid and the apex (the vertex opposite the base) a point *A* (freely chosen) in the lower base.

(2) Join the point *A* to the vertices *B, C, . . . , K* of the lower base.   Each segment so obtained is associated with a side of the upper base (an edge of the prismoid); the segment and the side form a pair of opposite edges of a tetrahedron.   (For instance, the segment *AB* is associated with the side *B′C′* and they determine together, as opposite edges, the tetrahedron *ABB′C′*.)

(3) The lines drawn from *A* to the vertices *B, C, . . . , K* dissect the lower base into triangles.   Each such triangle is associated with a vertex of the upper base; the triangle forms the base, the associated vertex the apex, of a pyramid (which is, in fact, a tetrahedron; for instance, *ABC* is associated with the vertex *C′*, and they determine together the pyramid *ABC-C′*).

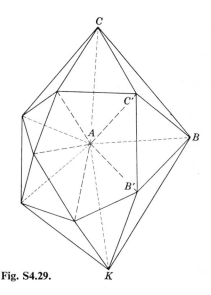

Fig. S4.29.

Our prismoid is dissected into the solids introduced in (1), (2), and (3).   (Of the upper base (1) takes the area, (2) the sides, (3) the vertices.   Of the lower base (1) takes just one point, (2) dividing segments, (3) the area.)   Apply ex. 4.26 to the pyramids (1) and (3) and ex. 4.28 to the tetrahedra (2).   Using ex. 4.27, you prove the prismoidal formula for the prismoid $BC \ldots KB'C' \ldots K'$ of Fig. S4.29.

4.30. The solution of ex. 4.28 is incomplete, since it treats only one out of three possible cases.   Consider two line segments: l and n, and the orthogonal projection n' of n on the plane parallel to n that passes through l.   Consider the two straight lines containing these two segments, respectively, and the point of intersection I of these two lines.   There are three possible situations: the point I may belong

(0) to none of the two segments l and n',
(1) to just one segment, or
(2) to both segments.

Ex. 4.20 treats only the case (2).   Yet a tetrahedron in situation (1) can be regarded as the difference of two tetrahedra in situation (2), and a tetrahedron in situation (0) can be regarded as the difference of two in situation (1).   In view of ex. 4.27, this remark completes the proof of ex. 4.28.

4.31. Fig. S4.29 is subject to two restrictions:
(1) Both bases are convex.
(2) Each vertex of one base is associated (there is a one-to-one correspondence) with a side of the other base: two lateral edges start from the vertex and end in the endpoints of the side.   (For instance, vertex B corresponds to the edge $B'C'$, vertex C' to the edge BC.)

Condition (2) is actually less restrictive than it might appear: many shapes which do not fall under it directly are limiting (degenerate) cases, and the proof extends to such shapes (by continuity, or appropriate interpretation).

The proof in ex. 4.35 is free from restrictions (1) and (2), but uses integral calculus.

4.32. $n = 0$; then $L = M = N = 1$, $I = 2$: valid.
$n = 2m - 1$, odd; $-L = N = 1$, $M = I = 0$: valid.
$n = 2m$, even; $L = N = 1$, $M = 0$, $I = 2/(n + 1)$:
valid for $n = 2$, but for no other positive even integer.

4.33. $f(x) = a + bx + cx^2 + dx^3$: superposition of the particular cases $n = 0, 1, 2, 3$ of ex. 4.32.

4.34. The substitution

$$x = a + \frac{h(t + 1)}{2}$$

transforms the interval $a \leq x \leq a + h$ into the interval $-1 \leq t \leq 1$ and any polynomial in x of degree ≤ 3 into another polynomial of the same kind in t.

4.35. We introduce a system of rectangular coordinates x, y, z.   We place the prismoid so that its lower base lies in the plane $z = 0$ and its upper base in the plane $z = h$. The volume of the prismoid is expressed by

$$(1) \qquad V = \int_0^h Q(z)\, dz$$

where $Q(z)$ denotes the area of the intersection of the prismoid with the plane parallel to, and at the distance z from, the lower base.

This intersection is a polygon with n sides if the prismoid has n lateral edges; its area is expressed by

$$(2) \qquad Q(z) = \tfrac{1}{2} \sum_{i=1}^{n} (x_i y_{i+1} - x_{i+1} y_i)$$

if the lateral edge number i is given by the pair of equations

$$(3) \qquad x_i = a_i z + c_i, \qquad y_i = b_i z + d_i$$

a_i, b_i, c_i, d_i are constants specifying the position of the edge; it is understood that edge number $n + 1$ coincides with edge number 1, so that

$$a_{n+1} = a_1, \qquad b_{n+1} = b_1, \ldots, y_{n+1} = y_1$$

Equations (2) and (3) show that $Q(z)$ is a *polynomial in z of degree not exceeding* 2, and so, by ex. 4.34, Simpson's rule, stated in ex. 4.32, is applicable to the integral (1), and this yields the prismoidal formula stated in ex. 4.22 since, obviously,

$$Q(0) = L, \qquad Q(h/2) = M, \qquad Q(h) = N$$

represent the areas of the lower base, the midsection and the upper base, respectively.

No solution: **4.11, 4.17, 4.36.**

Chapter 5

5.1. The unknown is the number V.

The data are the numbers a and h.

The condition is that V measures the volume of a right prism, the height of which is h and the base of which is a square with sides of length a.

5.2. There are two unknowns, the real numbers x and y. Or there is just one bipartite unknown, with components x and y, which we may interpret geometrically as a point in a plane with rectangular coordinates x and y.

The condition is fully expressed by the proposed equation.

We need not talk about data. (If we modified the problem by taking r^2 as the right-hand side of the proposed equation instead of 1, r would be a datum.)

A solution is $x = 1$, $y = 0$; another solution is $x = 3/5$, $y = -4/5$; and so on. In the geometrical interpretation, the full set of solutions consists of the points on the periphery of a circle with radius 1 and center at the origin.

5.3. There is no solution: the set of solutions is the empty set.

5.4. There are eight solutions:

$(2, 3)$ $(3, 2)$ $(-2, 3)$ $(-3, 2)$ $(2, -3)$ $(3, -2)$ $(-2, -3)$ $(-3, -2)$

The set consists of the lattice-points on the periphery of the circle with radius $\sqrt{13}$ and center at the origin. (A point of which both rectangular coordinates are integers is called a *lattice-point*. The configuration of lattice-points is important in number theory, crystallography, etc.)

5.5. We interpret the tripartite unknown (x, y, z) as a point in space with rectangular coordinates x, y, and z.

(1) The set of solutions consists of the points *in the interior* of an octahedron of which the center is at the origin and the six vertices are at the points

$(1, 0, 0)$ $(-1, 0, 0)$ $(0, 1, 0)$ $(0, -1, 0)$ $(0, 0, 1)$ $(0, 0, -1)$

(2) The set of solutions consists of the points in the interior and *on the surface* of the octahedron.

5.6. The following statement renders conspicuous the required principal parts:

If a, b, and c are the lengths of the sides of a right triangle, and a is the length of the side opposite the right angle

then $$a^2 = b^2 + c^2$$

5.7. We have to restate the theorem as the simultaneous assertion of two propositions in the usual "if-then" form where hypothesis and conclusion are conspicuous:

"If n is a square
then $d(n)$ is odd."
"If $d(n)$ is odd
then n is a square."

Here is a condensed statement, which uses "iff" (for "if and only if"):
"Iff n is a square then $d(n)$ is odd."

5.16. Begin by considering the case of a convex polygon and postpone whatever modifications may be needed to treat the general case.

(1) Given the lengths of $n - 1$ line-segments joining a chosen vertex of the polygon to the other $n - 1$ vertices, and the $n - 2$ angles included by pairs of consecutive line-segments.

(2) Divide the polygon by $n - 3$ diagonals into $n - 2$ triangles which are all determined (each by three sides) if the lengths of the dividing diagonals and of the n sides of the polygon are given.

(3) Take the particular case of the division into triangles considered under (2) supplied by the line-segments considered under (1). Number the triangles so that each triangle (except the first) has one side in common with the preceding triangle. Give any three independent data for the first triangle and, for each of the following $n - 3$ triangles, 2 data independent of each other and of the side that belongs also to the preceding triangle.

(4) Giving 2 rectangular coordinates for each of the n vertices, $2n$ data in

all, we would determine not only the polygon, but also a coordinate system the position of which is unessential. The position of the coordinate system depends on 3 parameters and so only $2n - 3$ data are essential.

5.17. To determine the base, $2n - 3$ data are required, see ex. 5.16. To determine the apex (the vertex opposite to the base) give the 3 rectangular coordinates of the apex with respect to a coordinate system of which a coordinate plane is the base, the origin a chosen vertex of the base, and a coordinate axis contains a side of the base starting from the chosen vertex. Hence, $2n$ data are required.

5.18. As in ex. 5.17, $2n$.

5.19. The polynomial is of the form

$$f_0 x_v^n + f_1 x_v^{n-1} + \cdots + f_{n-1} x_v + f_n$$

where f_j is a polynomial of degree j in $v - 1$ variables. Using ex. 3.34, we prove by mathematical induction that the number of data needed (the number of coefficients in the expansion in powers of $x_1, x_2, \ldots, x_v$) is

$$\binom{n + v}{v} = \binom{0 + v - 1}{v - 1} + \binom{1 + v - 1}{v - 1} + \cdots + \binom{n + v - 1}{v - 1}$$

No solution: **5.8, 5.9, 5.10, 5.11, 5.12, 5.13, 5.14, 5.15.**

Chapter 6

6.1. (1) 9 of shape $r(x) = 0$
(2) 9 of shape $r(x) = 0$
(3) 36 of shape $r(x, y) = 0$
(4) 7 of shape $r(x, y, z, w) = 0$
take into account obvious cancellations.
There are altogether 61 clauses.

6.2. Regard x as having the n components $x_1, x_2, \ldots, x_n$.

6.3. Let $x_1 = y_1$, $x_4 = y_3$, regard y_2 as having the two components x_2 and x_3, and regard the combination (the simultaneous imposition) of clauses r_2 and r_3 as one clause; then you have (with suitable notation) the "recursive" system

$$s_1(y_1) = 0$$
$$s_2(y_1, y_2) = 0$$
$$s_3(y_1, y_2, y_3) = 0$$

6.4. Regard y_1 as having the components x_1, x_2, x_3, and y_2 as having the components x_4, x_5, x_6, set $y_3 = x_7$, combine the first three clauses r_1, r_2, r_3 into one s_1, and the next three clauses r_4, r_5, r_6 into s_2: then you obtain the same final system as in ex. 6.3.

6.5. Essentially the same as the plan developed in sect. 6.4(2).

6.6. Particular case of the system in sect. 6.4(1). See ex. 3.21.

6.7. Two loci for a straight line; cf. sect. 6.2(5). In fact, all chords of given length in a given circle are tangent to an (easily constructible) circle concentric with the given circle.

6.8. Construct the point A on a, and the point B on b, each at the distance $l/2$ from the intersection of the lines a and b: draw the circle that touches a at A and b at B; since A may have two positions and the same is true of B, there are four such circles. One of these circles is an escribed circle of the desired triangle: the desired line x must be tangent to one of four circular arcs. We have here two loci for the straight line x; cf. sect. 6.2(5) and ex. 6.7.

6.9. HEARSAY.

6.11. (1) Begin with the "constant" c of the magic square. The sum of all nine unknowns, x_{ik} is, on the one hand,

$$= 1 + 2 + 3 + \cdots + 9 = 45$$

and it is, on the other hand, the sum of three rows and, therefore,

$$= 3c$$

From $45 = 3c$ follows $c = 15$.

(2) Add the three rows and the two diagonals; their sum is $5c$. Subtract hence those rows and columns that do *not* contain the central element x_{22}; their number is 4, their sum is $4c$. You obtain so

$$3x_{22} = 5c - 4c = 15$$

whence $x_{22} = 5$.

(3) In order to fill those rows and columns that do *not* contain the central x_{22}, list all different ways of representing 15 as the sum of three different numbers chosen among the following eight: 1, 2, 3, 4, 6, 7, 8, 9. A systematic survey yields:

$$15 = \mathbf{1} + 6 + 8$$
$$= 2 + 6 + \mathbf{7}$$
$$= 2 + 4 + \mathbf{9}$$
$$= \mathbf{3} + 4 + 8$$

(4) Those numbers that arise in *just one* of the above four representations of 15 are distinguished by heavy print; they must be placed in the middle of a row or column. The others (in ordinary print) arise *just twice*: they must be placed in a corner of the magic square.

(5) Start with any number in heavy print (for instance **1**) and set it $= x_{12}$. One of the numbers in ordinary print in the same representation of 15 (6 or 8 in our example) must be $= x_{11}$. The first time, you choose between 4 alternatives, the second time between 2, and you have no further choice: in using the four representations of 15 collected under (3), you can proceed further in just one way. You may obtain, for example, the magic square

6	1	8
7	5	3
2	9	4

and the $4 \times 2 = 8$ squares you can obtain are, in a sense, "congruent": you may derive all of them from any one of them by rotations and reflections. The square displayed shows the number 61 which arises in the solution of ex. 6.1.

6.12. (1) If the first digit of a number is ≥ 2, multiplication by 9 increases the number of digits. Therefore, the required number is of the form $1abc$.

(2) Moreover $1abc \times 9 = 9\ldots$, and so the number must be of the form $1ab9$.

(3) Therefore
$$(10^3 + 10^2 a + 10b + 9)9 = 9 \cdot 10^3 + 10^2 b + 10a + 1$$
$$89a + 8 = b$$

Hence, $a = 0$, $b = 8$: the required number is $1089 = 33^2$.

6.13. (1) Since $ab \times ba$ yields a three-digit number, $a \cdot b < 10$. Assume $a < b$; then there are only ten possible cases:
$$a = 1, \quad 2 \leq b \leq 9; \qquad a = 2, \quad b = 3 \text{ or } 4$$

(2) $(10a + b)(10b + a) = 100c + 10d + c$
$$10(a^2 + b^2 - d) = 101(c - ab)$$

Hence $a^2 + b^2 - d$ is divisible by 101; but
$$-9 < a^2 + b^2 - d \leq 82$$

Therefore, $a^2 + b^2 - d = 0$.

(3) $a^2 + b^2 = d \leq 9$. Hence $b < 3$, and so $a = 1$, $b = 2$; hence $c = 2$, $d = 5$.

6.14. (Mathematical Log, vol. II, no. 2.) Let us try to find such a paradoxical pair of triangles.

(1) Among those five parts there can not be three sides: otherwise the triangles would be congruent and all six parts would be identical.

(2) And so the triangles agree in two sides and three angles. Yet, if they have the same angles, they are similar.

(3) Let a, b, c be the sides of the first triangle, and b, c, d the sides of the second triangle: if they are, in this order, corresponding sides in our two similar triangles we must have
$$\frac{a}{b} = \frac{b}{c} = \frac{c}{d}$$

That is, the sides form a *geometric progression*. This can be done; here is an example:
$$a, \quad b, \quad c, \quad d$$
are equal to
$$8, \quad 12, \quad 18, \quad 27$$
respectively. Observe that $8 + 12 > 18$ and that the triangles, with sides 8, 12, 18, and 12, 18, 27 respectively, are similar, since their sides are proportional, and so they have the same angles.

6.15. (Mathematical Log, vol. III, no. 2 and 3.)

(*a*) Find three integers x, y, and z such that

$$x + y + z = 9, \qquad 1 \leqq x < y < z$$

A systematic survey yields just three solutions (just three ways of splitting 9 dollars);

$$9 = 1 + 2 + 6$$
$$= 1 + 3 + 5$$
$$= 2 + 3 + 4$$

(*b*) Arrange these three rows in a square so that also each column has the same sum 9.

Essentially (that is, except for permutations of the rows and the columns) there is just one such arrangement (which we present in a neat symmetric form):

$$\begin{array}{ccc} 6 & 2 & 1 \\ 2 & 4 & 3 \\ 1 & 3 & 5 \end{array}$$

(*c*) Now bring into play the remaining "minor" clauses of the condition: Since 6 is the greatest number in the square, the first row is for Al and the first column for ice cream. The only number in the square that equals twice the number in the intersection of the same row with the first column is 4; hence the second row is for Bill and the second column for sandwiches. And so Chris spent for soda pop the number in the intersection of the last row with the last column, 5 dollars.

6.16. (Mathematical Log, vol. III, no. 2 and 3)

(*a*) The wife buys x presents for x cents each, and the husband y presents for y cents each: The problem requires

$$x^2 - y^2 = 75$$

Now $75 = 3 \times 5 \times 5$ has just six divisors:

$$(x - y)(x + y) = 1 \times 75 = 3 \times 25 = 5 \times 15$$

and so there are just three alternatives:

$$x - y = 1 \qquad\qquad x - y = 3 \qquad\qquad x - y = 5$$
$$\text{or} \qquad\qquad\qquad \text{or}$$
$$x + y = 75 \qquad\qquad x + y = 25 \qquad\qquad x + y = 15$$

which yield the table:

wife	husband
38	37
14	11
10	5

(*b*) Now bring into play the remaining "minor" clauses of the condition. They show *unambiguously*

Ann	38	37	Bill Brown
	14	11	Joe Jones
Betty	10	5	

and so Mary's last name *must* be Jones.

6.17. (Cf. Archimedes, vol. 12, 1960, p. 91.)   Obviously, the number of cases is restricted from the start ($4! = 24$).   Yet, if you are smart, you need not examine all these cases.

(*a*) Let

$$b, \qquad\qquad g, \qquad\qquad w, \qquad\qquad s,$$

stand for the number of bottles consumed by the wife of

$$\text{Brown,} \qquad \text{Green,} \qquad \text{White,} \qquad \text{Smith,}$$

respectively.   Then

$$b + g + w + s = 14$$
$$b + 2g + 3w + 4s = 30$$

and so

$$g + 2w + 3s = 16$$

(*b*) As the last equation shows, either g and s are both odd or they are both even.   Hence, there are only 4 cases that need to be examined:

g	s	$w = 8 - (g + 3s)/2$
3	5	-1
5	3	1
2	4	1
4	2	3

Only the last case is admissible.   Therefore

$$s = 2, \qquad w = 3, \qquad g = 4, \qquad b = 5$$

and the ladies are

Ann Smith,   Betty White,   Carol Green,   Dorothy Brown.

6.18. The division of the condition into clauses is often useful in solving puzzles.   The reader may find appropriate examples in collections of mathematical puzzles, for instance in H. E. Dudeney, *Amusements in Mathematics* (Dover).   The *Otto Dunkel Memorial Problem Book* published as supplement to the *American Mathematical Monthly* **64** (1957) contains some suitable material; no. E776 on p. 61 deserves to be quoted as an exceptionally neat example of its type.

6.20. (*b*) sect. 6.2(4), $l = 5$.   (*c*) ex. 6.6, $n = 4$.   (*d*) sect. 6.4(1), $n = 4$. (*e*) sect. 2.5(3), sect. 6.4(2).

6.22. In this system of three equations each of the three unknowns x, y, and z plays exactly the same role: A cyclical permutation of x, y, and z interchanges the 3 equations, but leaves their system unchanged.   Therefore, *if* the unknowns are uniquely determined, we must have $x = y = z$; supposing this, we have immediately $6x = 30$, $x = y = z = 5$.

There remains to prove that the unknowns are uniquely determined; this can be shown by some usual procedure to solve a system of linear equations.

6.23.
$$y + z = a$$
$$x + z = b$$
$$x + y = c$$

Any permutation of x, y, and z leaves the system of the left-hand sides unchanged. Set $x + y + z = s$ (which also remains unchanged by a permutation of x, y, and z); adding the three equations we easily find

$$s = (a + b + c)/2$$

and the system reduces to three equations, each containing just one unknown

$$s - x = a, \qquad s - y = b, \qquad s - z = c$$

The whole system (not only the left-hand sides) is symmetric with respect to the pairs (x, a), (y, b), (z, c).

6.24. (Stanford 1958.) Set

$$x + y + u + v = s$$

and see ex. 6.23.

No solution: **6.10, 6.19, 6.21, 6.25.**

APPENDIX

Hints to Teachers, and to Teachers of Teachers

Teachers who wish to make use of this book in their profession should not neglect the hints addressed to all readers, but they should pay attention also to the following remarks.

1. As explained in the Preface, this book is designed to give opportunity for *creative work on an appropriate level* to prospective high school mathematics teachers (also to teachers already in service). Such opportunity is desirable, I think: a teacher who has had no personal experience of some sort of creative work can scarcely expect to be able to inspire, to lead, to help, or even to recognize the creative activity of his students.

The average teacher cannot be expected to do research on some very advanced subject. Yet the solution of a nonroutine mathematical problem is genuine creative work. The problems proposed in this book (which are not marked with a dagger) do not require much knowledge beyond the high school level, but they do require some degree, and sometimes a high degree, of concentration and judgment. The solution of problems of this kind is, in my opinion, the kind of creative work that ought to be introduced into the high school mathematics teachers' curriculum. In fact, in solving this kind of problems, the prospective teacher has an opportunity to acquire *thorough knowledge of high school mathematics*—real knowledge, ready to use, not acquired by mere memorizing but by applying it to interesting problems. Then, which is even more important, he may acquire some *know-how*, some skill in handling high school mathematics, some insight into the essentials of problem-solving. All this will enable him to lead, and to judge, his students' work more efficiently.

2. The present volume contains only the first half of a full course. Especially, teaching methods are only implicitly suggested by this volume; they will be explicitly discussed in a chapter of the second volume.

Yet this first volume contains much problem material that could be used in certain (especially, in more advanced) high school courses. I propose as a useful exercise to teachers to *reflect on possible classroom use* of the problems they are doing.

The best time for such reflection may be when the solution has been obtained and well digested. Then you *look back* at your problem and ask yourself: "Where could I use this problem? How much previous knowledge is needed? Which other problems should be treated first to prepare the class for this one? How could I present this problem? How could I present it (be specific) to such and such a class—or how could I present it to Jimmy Jones?" All these questions are good questions and there are many other good questions—but the best question is the one that comes spontaneously to your mind.

3. Although this first volume does not present a full course, it contains enough material to serve as textbook for a *Seminar in Problem Solving*. I conducted such seminars in various Institutes for teachers; several colleagues interested in starting such a seminar asked me for my materials; I know of a few colleges where such seminars, or similar classes, have been actually offered lately; and it is, I think, highly desirable that many more colleges should start experimenting with such seminars. It is in view of this situation that I have decided to publish this first volume before the second, in spite of the obvious risks of such an incomplete publication.

4. After some trials, I worked out a procedure for my seminar, a description of which at this place may be useful.[1]

Typical problems, which indicate a useful pattern, are solved in class discussion led by the instructor; the text of the first four chapters reproduces (as closely as it can be done in print) such class discussions. Then the discussion leads to recognizing and formulating the pattern involved—the text of the chapters quoted shows also how this is done.

The homework of the students consists of problems (such as the problems printed here following each chapter) which offer an opportunity to apply, to clarify, and to amplify the pattern obtained (and also the methodical remarks made) in class.

5. I used my seminar (and this is an essential feature of it) to give the participants some practice in explaining problems and guiding their solu-

---

[1] Some of the following will be, and a few foregoing sentences have been, extracted from an article in *The Journal of Education, Vancouver and Victoria*; see the Bibliography.

tion, in fact, some opportunity for *practice teaching*, for which in most of the usual curricula there is not enough opportunity.

When the homework is returned, this or that point (a more original solution, a more touchy problem) is presented to the class on the blackboard by one of the participants who did that point particularly well, or particularly badly. Later, when the class has become more familiar with the style of the performance, a participant takes for a while the instructor's place in leading the class discussion. Yet the best practice is offered in *group work*. This is done in three steps.

First, at the beginning of a certain practice session, each participant receives a different problem (each just one problem) which he is supposed to solve in that session; he is not supposed to communicate with his comrades, but he may receive some help from the instructor.

Then, between this session and the next, each participant should check, complete, review, and, if possible, simplify his solution, look out for some other approach to the result and, by these means and any other means, master the problem as fully as he can. He should also do some planning for presenting his problem and its solution to a class. He is given opportunity to consult the instructor about any of the above points.

Finally, in the next practice session, the participants form *discussion groups*: each group consists of four members (there may be one odd group); the participants form these groups by mutual consent, without intervention of the instructor. One member of the group takes the role of the teacher, the other members act as students. The "teacher" presents his problem to the "students," tries to challenge their initiative and tries to guide them to the solution, in the same style as the instructor does it in class discussions. When the solution has been obtained, a short friendly criticism of the presentation follows. Then another member takes the role of the teacher and presents his problem, and the procedure is repeated until each member of the group has had his turn. Then the participants partially regroup (each of two neighboring groups may send a member as "teacher" to the other group) so that each participant has occasion to polish his performance in presenting his problem several times. Some particularly interesting problems or particularly good presentations are shown to, and afterwards discussed by, the whole class. Congenial groups may spontaneously undertake the discussion of problems which are new to all participants; this should be encouraged, of course.

Such problem solving by discussion groups became quickly very popular in my classes, and I have the impression that the seminars as a whole were a success. Many of the participants were experienced teachers and several of them felt that their participation suggested to them useful ideas about conducting their own classes.

6. This volume may help the college instructor who conducts a Seminar in Problem Solving (especially when he conducts it for the first time). He may follow the procedure just described (in sect. 4 and 5). In class discussions, he may use the text of any one of the four first chapters. The problems printed at the end of the chapter are suitable for homework: serious work may be needed to expand the sketch of a solution given at the end of the volume into a fully presented solution. (Yet the instructor cannot choose at random: he should have a good look at the problem, at its solution, and also at the problems surrounding it, before assigning it.) For examinations and term papers, the instructor may wish to avoid problems printed in this volume; then he may consult appropriate textbooks (and also ex. 1.50, 2.78, and 3.92). For group work (see sect. 5) the problems should be harder, but they need not be closely connected with the chapters treated; suitable problems may be selected from this book, also from later chapters.

Chapters 5 and 6 may be also discussed, or assigned for reading. The role of these chapters, however, can be better explained in the second volume.

Of course, after having gained some experience, the instructor may adopt the spirit of this book without following its details too closely.

BIBLIOGRAPHY

I. Classics

EUCLID, *Elements*. The inexpensive shortened edition in Everyman's Library is sufficient here. ("Euclid III 20" refers to Proposition 20 of Book III of the Elements.)

PAPPUS ALEXANDRINUS, *Collectio*, edited by F. Hultsch, 1877; see v. 2, pp. 634–637 (the beginning of Book VII).

DESCARTES, *Œuvres*, edited by Charles Adam and Paul Tannery. (For remarks on the "Rules"—the work which is of especial interest for us—and the way of quoting it, see ex. 2.72.)

LEIBNITZ (or LEIBNIZ) (1) *Mathematische Schriften*, edited by C. J. Gerhardt. (2) *Philosophische Schriften*, edited by C. J. Gerhardt. (3) *Opuscules et fragments inédits*, collected by Louis Couturat.

BERNARD BOLZANO, *Wissenschaftslehre*, second edition, 1930; see v. 3, pp. 293–575 (Erfindungskunst).

II. More modern

E. MACH, *Erkenntnis und Irrtum*, fourth edition, Leipzig, 1924; see pp. 251–274 and *passim*.

J. HADAMARD, *Leçons de Géométrie plane*, Paris, 1898; voir Note A, *Sur la méthode en géométrie*.

F. KRAUSS, Denkform mathematischer Beweisführung, *Zeitschrift für mathematischen und naturwissenschaftlichen Unterricht*, v. 63, pp. 209–222.

WERNER HARTKOPF, *Die Strukturformen der Probleme*; Dissertation, Berlin, 1958.

III. Related work of the author

Books

1. *Aufgaben und Lehrsätze aus der Analysis*, 2 volumes; third revised edition, Berlin, 1964; jointly with G. SZEGÖ.

2. *How to Solve It*; second edition, 1957; Anchor Book A 93, Doubleday. (Quoted as HSI; references are to the pages of the second edition but, for the convenience of the users of former printings, titles are added.)
3. *Mathematics and Plausible Reasoning*, Princeton, 1954. In two volumes, entitled *Induction and Analogy in Mathematics* (vol. I) and *Patterns of Plausible Inference* (vol. II). (Quoted as MPR.)

Papers

1. Geometrische Darstellung einer Gedankenkette. *Schweizerische Pädagogische Zeitschrift*, 1919, 11 pp.
3. Wie sucht man die Lösung mathematischer Aufgaben? *Acta Psychologica*, **4**, 1938, pp. 113–170.
11. Die Mathematik als Schule des plausiblen Schliessens. *Gymnasium Helveticum*, **10**, 1956, pp. 4–8; reprinted *Archimedes*, **8**, 1956, pp. 111–114. Mathematics as a subject for learning plausible reasoning; translation by C.M. Larsen. *The Mathematics Teacher*, **52**, 1959, pp. 7–9.
12. On picture-writing. *American Mathematical Monthly*, **63**, 1956, pp. 689–697.
13. L'Heuristique est-elle un sujet d'étude raisonnable? *La Méthode dans les Sciences Modernes* ("Travail et Méthode," numéro hors série) 1958, pp. 279–285.
14. On the curriculum for prospective high school teachers. *American Mathematical Monthly*, **65**, 1958, pp. 101–104.
15. Ten Commandments for Teachers. *Journal of Education of the Faculty and College of Education, Vancouver and Victoria*, nr. 3, 1959, pp. 61–69.
16. Heuristic reasoning in the theory of numbers. *American Mathematical Monthly*, **66**, 1959, pp. 375–384.
17. Teaching of Mathematics in Switzerland. *American Mathematical Monthly*, **67**, 1960, pp. 907–914; *The Mathematics Teacher*, **53**, 1960, pp. 552–558.
18. The minimum fraction of the popular vote that can elect the President of the United States. *The Mathematics Teacher*, **54**, 1961, pp. 130–133.
(The numbers omitted are quoted in MPR, v. 1, pp. 279–280, and v. 2, pp. 189–190.)

IV. Problems

Among the examples proposed for solution, some are taken from the *Stanford University (Stanford-Sylvania) Competitive Examination in Mathematics*. This fact is indicated at the beginning of the solution with the year in which the problem was given as "Stanford 1957." Most of these problems (some with solution) appeared in *The California Mathematics Council Bulletin*.

The *Olympiad Problem Book*, by Shklarsky, Chentzov, and Yaglom, contains many unusual and difficult elementary problems proposed in Russian competitive examinations. An English translation, by *I. Sussman*, is scheduled to be published by W. H. Freeman and Co.

INDEX